Civil Society and Social Movements in Food System Governance

This book offers insights into the governance of contemporary food systems and their ongoing transformation by social movements.

As global food systems face multiple threats and challenges there is an opportunity for social movements and civil society to play a more active role in building social justice and ecological sustainability. Drawing on case studies from Canada, the United States, Europe, and New Zealand, this edited collection showcases promising ways forward for civil society actors to engage in governance. The authors address topics including: the variety of forms that governance engagement takes from multi-stakeholderism to co-governance to polycentrism/self-governance; the values and power dynamics that underpin these different types of governance processes; effective approaches for achieving desired values and goals; and, the broader relationships and networks that may be activated to support change. By examining and comparing a variety of governance innovations, at a range of scales, the book offers insights for those considering contemporary food systems and their ongoing transformation.

It is suitable for food studies students and researchers within geography, environmental studies, anthropology, policy studies, planning, health sciences and sociology, and will also be of interest to policy makers and civil society organizations with a focus on food systems.

Peter Andrée is Associate Professor in Political Science, Geography and Environmental Studies at the Institute of Political Economy at Carleton University, Canada.

Jill K. Clark is Associate Professor at the John Glenn College of Public Affairs at Ohio State University, USA.

Charles Z. Levkoe is Canada Research Chair in Sustainable Food Systems and Assistant Professor in the Department of Health Sciences, Lakehead University, Canada.

Kristen Lowitt is Assistant Professor in the Department of Geography, Brandon University, Canada.

Routledge Studies in Food, Society and the Environment

Food and Nutrition Security in Southern African Cities
Edited by Bruce Frayne, Jonathan Crush and Cameron McCordic

The Real Cost of Cheap Food
(Second Edition)
Michael Carolan

Food Bank Nations
Poverty, Corporate Charity and the Right to Food
Graham Riches

Urban Food Systems Governance and Poverty in African Cities
Edited by Jane Battersby and Vanessa Watson

Risk and Food Safety in China and Japan
Theoretical Perspectives and Empirical Insights
Edited by Louis Augustin-Jean and Jean-Pierre Poulain

Localizing Global Food
Short Food Supply Chains as Responses to Agri-Food System Challenges
Edited by Sophia Skordili and Agni Kalfagianni

Seafood Supply Chains
Governance, Power and Regulation
Miriam Greenwood

Civil Society and Social Movements in Food System Governance
Edited by Peter Andrée, Jill K. Clark, Charles Z. Levkoe and Kristen Lowitt

For further details please visit the series page on the Routledge website: http://www.routledge.com/books/series/RSFSE/

Civil Society and Social Movements in Food System Governance

Edited by Peter Andrée, Jill K. Clark, Charles Z. Levkoe and Kristen Lowitt

Routledge
Taylor & Francis Group
LONDON AND NEW YORK

earthscan
from Routledge

First published 2019
by Routledge

2 Park Square, Milton Park, Abingdon, Oxfordshire OX14 4RN
52 Vanderbilt Avenue, New York, NY 10017

Routledge is an imprint of the Taylor & Francis Group, an informa business

First issued in paperback 2020

British Library Cataloguing-in-Publication Data
A catalogue record for this book is available from the British Library

Library of Congress Cataloging-in-Publication Data
A catalog record has been requested for this book

ISBN: 978-1-138-58807-3 (hbk)
ISBN: 978-0-367-66182-3 (pbk)

Typeset in Sabon
by Swales & Willis Ltd, Exeter, Devon, UK

Contents

Acknowledgements

This book has solid footing in the work of the Food: Locally Embedded, Globally Engaged (FLEdGE) action research collaborative. FLEdGE is a research and knowledge sharing partnership committed to fostering food systems that are socially just, ecologically regenerative, economically localized, and that engage a wide range of people across food systems. Specifically, we want to acknowledge the 86 partners, co-applicants, and collaborators in Canada and 25 collaborators in the international community that are on the front lines of this work and have inspired this project. FLEdGE receives funding from the Social Sciences and Humanities Research Council of Canada (SSHRC).

Most of the chapters in this volume were first presented in a workshop held in conjunction with a FLEdGE meeting in September 2017. We greatly appreciate the participation of fellow researchers and community partners, which included reviewing and discussing early ideas and drafts of chapters. We specifically want to thank the following individuals: Elizabeth Bacon, Patricia Ballamingie, Jane Battersby, Mary Coulas, Ana Moragues Faus, Harriet Friedmann, Tim MacNeill, Hugo Martorell, Ken Meter, and Theresa Schumilas.

Projects like these do not happen without the help of colleagues and friends. Foremost, we thank Alison Blay-Palmer who played an instrumental role in bringing together this community-based research and practice community and providing (and encouraging us!) with the forum and resources to bring this book to fruition. The feedback from Molly Anderson and Phil Mount on the theoretical framework presented in Chapter 1 was critical to its development. A special thank you to Carla Johnston for her careful and consistent attention to details when editing the volume. We greatly appreciate the support from Tim Hardwick and Amy Johnston from Routledge/Taylor & Francis Group. This work has been published open access thanks to grants from the Aid to Publication Program of Saint Paul University, the Faculty of Extension, University of Alberta, the Department of Geography, University of Otago, the Food Studies Program, Middlebury College, the John Glenn College of Public Affairs, Ohio State University, Carleton University, and the Social Sciences and Humanities Research Council (SSHRC), Wilfrid Laurier University and FLEdGE.

Finally, we hope that researchers, policy makers and food movement activists find the work presented in this book useful.

Contributors

Peter Andrée is Associate Professor and Associate Chair in the Department of Political Science at Carleton University. His research focuses on the politics of food systems and the environment. He practices, and teaches, community-based participatory research methods. Prof. Andrée is co-editor of *Globalization and Food Sovereignty: Global and Local Change in the New Politics of Food* (2014) and author of *Genetically Modified Diplomacy* (2007).

Molly D. Anderson is the William R. Kenan, Jr. Professor of Food Studies at Middlebury College and Academic Director of its Food Studies Program. She worked previously at Oxfam America and Tufts University, and has published widely on sustainable food systems issues.

Elizabeth Bacon is a research assistant with Dr. Mary Beckie at the University of Alberta, as part of the FLEdGE network. In projects related to FLEdGE, she has had the opportunity to investigate local food policies and food system governance in Alberta. Currently, Elizabeth works at the Secretariat of the Convention on Biological Diversity in Montreal.

Mary Beckie is an Associate Professor and Director of Community Engagement Studies at the University of Alberta. Her research on sustainable and localized agri-food systems has taken place in western Canada, the European Union, Cuba, India and Sri Lanka. Mary is the academic lead for the Alberta/British Columbia node of the FLEdGE research partnership.

Jill K. Clark is an Associate Professor in the John Glenn College of Public Affairs at Ohio State University. Her research interests include food policy and practice, centering on community and state governance of food systems, the policy process, and public participation.

Chantal Wei-Ying Clément is Coordinator of the International Panel of Experts on Sustainable Food Systems (IPES-Food). She holds a Ph.D. in Political Science from Carleton University.

Sean Connelly is a Senior Lecturer in the Department of Geography at the University of Otago. He teaches courses on environmental management and his research interests are in the broad area of sustainable communities, with particular focus on alternative food systems and rural and regional development.

Carla Johnston is a Ph.D. Candidate and a Doctoral Fellow at the Balsillie School of International Affairs. Her research interests include the governance of sustainable food systems in northern Canada as well as using Participatory Action Research (PAR) methodology to work directly with civil society groups to create meaningful actions that help them reach their goals.

Ryan Lauzon is at the Fisheries Assessment Program, Chippewas of Nawash Unceded First Nation. Ryan is responsible for supporting informed fisheries management decisions at the Saugeen Ojibway Nation Joint Council, through the collection of data on the commercial fishery.

Charles Z. Levkoe is the Canada Research Chair in Sustainable Food Systems in the Department of Health Sciences at Lakehead University. Charles' community-engaged research uses a food systems lens to explore connections between social justice, ecological regeneration, regional economies, and democratic engagement.

Kristen Lowitt is at the Department of Geography, Brandon University, Canada. Her research looks at the interactions among food security, communities, and natural resource management in rural and remote regions.

Philippa MacKay is an Environmental Consultant who works on strategic and community development projects. Her passion lies with the implementation of sustainable practice more generally, but she has a keen interest in sustainable food system change. She completed her Masters of Planning in the Department of Geography at the University of Otago.

Phil Mount is the Associate Director of Just Food in Ottawa, an Associate Researcher at the Laurier Centre for Sustainable Food Systems, and Board Chair of Sustain Ontario. Phil was born and raised on a dairy farm in Bearbrook, and now digs and levels with his wife Denise, building Flat Earth Farm in rural south Ottawa.

Kathleen Ryan is at the Saugeen Ojibway Nation Environment Office, Canada, Kathleen Ryan (Anungkwe) holds a BSc. in Indigenous Science and an MSc. in Aquatic Ecology. Kathleen works at the Saugeen Ojibway Nation Environment Office in Neyaashiinigmiing.

Dean Sayers has been Chief of Batchewana First Nation since 2005. He grew up in Batchewana village, a small community north of Sault Ste. Marie, Ontario, where he worked with his father and brother in the First Nation's Commercial Fishing industry. Chief Sayers' experiences and historical understanding of Batchewana and its people have been instrumental to the First Nation's success and assertions of sovereignty and jurisdiction over their land and resources.

Amanda Wilson is an Assistant Professor in the School of Innovation at Saint Paul University in Ottawa, Canada. Her research is focused on food movements and alternative food networks, cooperatives, and collective organizing, and questions related to prefiguration and enacting a politics of possibility.

Foreword

Social movements in food system governance

Phil Mount

The first two decades of the 21st century have produced ominous disruptions of millenarian portent. The reverberations of 9/11 and the Great Recession continue to shape global geo-political and socio-economic realities, magnifying the effects of migration, forced relocation and social unrest, decades of rampant trade liberalization, and ever-increasing disparities in wealth—within and between communities and countries.

The fire under this bubbling cauldron draws fuel from the normalization of identity politics and rhetorical bombast; the exponential growth of celebrity and social media as the filters through which we consume information; and the deterioration and opinionification of social discourse—resulting in a politics of polarization, and a polarization of politics.

The new spaces in the realm of public discourse are being claimed by a diverse set of actors no longer marginalized by conventional standards of interaction, communications, and governance. The Age of Disruption has, at the same time, spawned ISIL, Brexit, TPP and Trump, Arab Spring, Wikileaks, Occupy, Idle No More, Black Lives Matter, and #Me2.

Similarly, there is no doubt that this same fuel has ignited dissatisfaction with the impacts of the status quo—for farmers, food insecure, public health, the environment, and rural communities—to produce disruptions in our food systems, disruptions that ripple from local to global and back. La Via Campesina (LVC) began as a global network of grassroots actors, raising the flag of food sovereignty in the face of trade liberalization. As a movement, LVC picked up strength by connecting under that banner organizations with diverse interests and/or complaints—including agricultural land grabbing by state and private financial interests; gender parity and an end to violence against women; the rights of indigenous, informal sector, and agricultural workers; and increasing privatization, concentration, and corporate domination from seed to fork (of which the Bayer/Monsanto merger is only the latest and most egregious example).

Many of the same actors have coalesced to lead an agroecology movement which integrates food sovereignty within a framework that foregrounds ecological and food justice goals, espousing an alternative vision with a food systems approach that pays equal attention to ecological regeneration, producer livelihoods, community well-being, and food justice. This movement has gained the attention of global food power players, and traction within the Food and Agriculture Organization of the United Nations, leading to the dilemma of social movement politics:

> Do we engage in a forum where our message may be distorted and/or co-opted, in exchange for a platform offering a chance to influence key decision-makers, and drive systems transformation?

This question—at the intersection of social movements, systems thinking, and governance—is at the heart of this book, which showcases through diverse examples how, why, and with what caveats food movement actors are becoming actively involved in decision-making at local, regional, and international levels.

In fact, it is interesting to note that many of these examples focus on the local or regional level, where the devolution (or abdication) of state responsibility under neoliberalism opens space to do things differently, with different 'stakeholders', for different reasons. As an academic, the possibility of contributing to these efforts is what first attracted me to community-based research and work on a regional food strategy. My engagement with the Food: Locally Embedded, Globally Engaged project over the last four years has included many private discussions and public roundtables with academic and activist colleagues, on the merits and challenges of engaging in food systems governance processes. This has demonstrated to me more clearly than ever the value of the academic-activist connection, and the lessons we can learn from each other as we engage in these complicated governance conversations.

Now that I am also working at the community level in a grassroots food systems organization, engagement with food systems governance processes seems less like an optional public service. Resource constraints (staff, time, and funding) mean such organizations have to maximize their reach and impact through collaboration in spaces that deliver tangible, salient results through a consensus of diverse actors. Small organizations that use this approach consistently punch above their weight class.

However, food systems governance processes invariably involve trade-offs and uncomfortable conversations—not necessarily familiar territory for social movement actors. The lessons in these pages—from social movement actors who are starting to chart this territory—will help all of us in our efforts to turn disruption into transformation.

Introduction – Traversing theory and practice

Social movement engagement in food systems governance for sustainability, justice, and democracy

Peter Andrée, Jill K. Clark, Charles Z. Levkoe, and Kristen Lowitt

Introduction

Globally, food systems are in a state of flux in response to a wide range of forces, including consumer and peasant movement demands, the (re)negotiation of trade agreements, and technological changes in agriculture production, processing, and distribution (Clapp, 2016; Andrée et al, 2014). The instability of the dominant food system, premised on industrial methods and corporate control, is also affected by the political imperative to respond to a complex set of issues, including the challenges resulting from the financialization of food and the volatility of the global marketplace, climate change mitigation and adaptation, food access and safety, and diet-related diseases. This state of flux represents a critical historical moment, full of both challenge and opportunity, for social movements organizing around food to build a more sustainable and just world (Levkoe, 2014; Wittman et al, 2011).

The past two decades have seen an uprising of movements that challenge industrial food systems by experimenting with a variety of alternative ways of producing, harvesting, foraging, processing, distributing, consuming, and, ultimately, governing food. These movements seek to reinforce, build on, and scale-up innovative, place-based initiatives. These initiatives consider food as part of an interrelated system that includes not just its quality and health, but also the broader context of how food moves from the fields, water, and forests to our plates and beyond. These movements, and the initiatives they spearhead, are associated with a range of labels including fair trade, civic agriculture, food justice, food sovereignty, food democracy, agroecology, slow food, and community food security (Schiff & Levkoe, 2014; Friedland, 2010; Hendrickson & Heffernan, 2002). We refer to this collection of actors with related objectives as "food movements", in the plural. This appellation recognizes their diversity while acknowledging overlapping and, at times, conflicting goals.[1] One reason to refer to food movements collectively is because, as social movements, they are increasingly connected. For example, Levkoe (2015) describes Canada's food movement as a "network of networks", highlighting the collaborations across sectors, scales, and places found in contemporary food activism. This connectivity relates to the growing tendency of food movement actors to adopt an integrated, system-wide perspective that seeks to address multiple elements of food systems, including how decisions are made, and by whom.

Two examples from different scales illustrate the increasing connectivity, shared analysis, and shared goals of today's food movements. In Canada, the People's Food Policy (PFP) project launched a ground-breaking report in 2011 that synthesized food system-related concerns and aspirations from a variety of social movements across the country, including activists from farmers' unions, Indigenous groups, international development organizations, urban food security organizations, labour unions, environmental groups, and others (Levkoe & Sheedy, 2018). The result of their conversations, the *Resetting the Table* report, argued that to transform food systems, work was needed to simultaneously respect Indigenous food sovereignty, develop more sustainable production systems, eliminate poverty, and ensure access to sustainably-produced food for all (PFP, 2011). One of the ten working papers produced by the PFP that focused on food democracy and governance also put forward a political argument, stating that democratic societies must "guarantee the meaningful and active involvement of all individuals, groups, and institutions in decision-making processes" (Levkoe et al, 2011, p. 5). In a second example, reports of the International Panel of Experts on Sustainable Food Systems (IPES-Food) have adopted a similar systems-wide perspective. Established only in 2015, but rapidly gaining an international audience and policy influence, IPES-Food seeks to diagnose the "lock-ins" of the industrial food system and identify pathways for (re)building alternatives rooted in food sovereignty and agroecology. Along with the PFP in Canada, IPES-Food argues both for transforming food systems and for reimagining who is included in the decisions about the shape of those food systems (IPES, 2015).

The growing connection among various food movements has come about, in part, as a response to calls like that of Amin (2011) for food movements to unite while "recognizing the diversity, not only of movements, which are fragmented, but of the political forces that are operating with them, of the ideologies and even visions of the future of those political forces" (p. xvii). Another reason for this connectivity is hinted at in Amin's call: Beyond focusing on place-based initiatives, many of these efforts involve scaling-up local knowledge and experience, and collaborating with others along the way, with an aim to exact a political impact on the structures of food systems (Schiff & Levkoe, 2014). To effect such change, food movement actors increasingly recognize the need, in addition to working outside of formal governance processes as critics, to become agents of systemic reform in order to scale up, and scale out, alternative food system possibilities (Blay-Palmer et al, 2016).

As the chapters in this book demonstrate, an increasing number of organizations associated or allied with food movements are taking this step and actively engaging in food systems governance. Their engagement takes a variety of forms. Some organizations seek to influence government policy while others work to influence particular stakeholders (e.g. industry policies and practices). Still others are actively taking on collaborative governance (or co-governance) roles themselves. Co-governance means they have an equal role alongside other actors (whether formal governments or the private sector) as co-producers of governance outcomes. In light of these developments, important questions emerge: What does the increasing engagement of food movement actors in governance look like in relation to the power dynamics and spatial organization of food systems? What are the implications of different forms of governance relationships on the effectiveness of addressing social movement goals? To what extent do these relationships affect the transformative potential of food systems? What are the values that underpin various governance arrangements and how are these negotiated? How do co-governance arrangements, in particular, work in practice and what are their effects? How does framing and opportunity vary across scales? Further, from the perspective of movement actors: What strategies are most effective for achieving desired values and goals in specific contexts? What are the risks and costs associated with active participation in co-governance regimes?

These are all questions that *Civil Society and Social Movements in Food System Governance* seeks to address. The overall purpose of this collection is to examine the opportunities, challenges, and implications of a variety of context-specific governance interventions by food movement actors at a range of food systems scales. While many critical scholars focus their energies on identifying what is wrong with the governance of food systems (e.g. Clapp, 2016; Howard, 2016), this effort aims to identify emerging opportunities for harnessing governance processes to build healthy, equitable, and sustainable food for all. We seek to articulate and provide visibility to promising ways forward for social movement engagement in food systems governance at multiple levels. The authors in this collection came together through a governance working group as part of the Food: Locally Embedded Globally Engaged (FLEdGE) network. FLEdGE is a community-engaged research partnership that explores the current and potential role of community food initiatives to act as pillars of regional, sustainable food systems. The book's authors represent a range of interdisciplinary scholars working collaboratively with civil society partner organizations on food systems issues across the Global North.[2] Together, we are exploring how governance innovations can help to catalyze, integrate, or support broader progressive economic, political, and/or cultural transformations.

This collection builds on FLEdGE's history of community-academic engagement as well as recent books such as *Food Systems Governance* (Kennedy & Liljeblad, 2016) and *Public Policies for Food Sovereignty* (Desmarais et al, 2017).

Its distinct contribution is to examine the opportunities, challenges, and implications of food movements' engagement with governance, with a particular attention to co-governance. While the editors are sympathetic towards food systems initiatives rooted in social and ecological justice, this collection does not focus on the achievements associated with one specific ideological frame (e.g. food sovereignty) or operational model (e.g. cooperatives, food policy councils, Right to Food legislation). Rather, by bringing together case studies that reach across a range of food movements, we are able to glean lessons more generally about the possibilities and challenges of governance engagement, and to describe some of the opportunities, limitations, and gaps that point towards a new set of research questions and that may help inform the practice of our social movement partners.

The collection of chapters in this volume demonstrate: 1) the central importance that engagement in governance processes is taking on for food movements seeking to have a wider, systemic, impact; 2) the variety of scales of governance that food movements are engaging with, though there is an emphasis on the local scale, as this appears to be the entry point for movements making the shift from food systems initiatives to more direct governance; 3) the variety of forms that governance engagement takes along a continuum, from multi-stakeholderism to co-governance to polycentrism/self-governance; 4) the central role of building relationships with other actors based on mutual trust and commitment to change while moving along the governance continuum. The continuum enables discussion of the benefits and challenges associated with those different levels of engagement including the compromises that specific forms of engagement can lead to. This collection also demonstrates the important role that building trust with other actors plays for those who seek to move along the governance continuum from multi-stakeholderism to co-governance.

The remainder of this introductory chapter provides context for the case studies in this collection. It begins with a definition of governance, followed by a discussion of the relationship between transitions taking place in food systems governance processes and neoliberalism. Neoliberalism is foregrounded because, along with associated practices, policies, and processes, it has enabled new governance arrangements. At the same time, neoliberal processes have served to undermine protections of equity, sustainability, and democracy that many of these new governance mechanisms now seek to ensure. We then define our use of the terms civil society and social movements in this book, followed by a discussion of how we might start to conceptualize the impacts of food movements, in particular, on food systems governance. We finish by presenting an overview of the book's layout and the key contributions of each chapter.

Transitions in food systems governance

At the heart of this collection is the idea of governance. By governance, we are referring to the relationships, processes, rules, practices, and structures (both institutional and discursive) through which power and control are exercised and decisions are made. Governance processes include more than the state and its associated agencies, given the formal and informal roles of a wide range of actors in policy-making processes (Minnery, 2007; Jessop, 2002). The concept of governance takes us beyond "government" in two ways. First, it acknowledges that there are always multiple actors and perspectives involved in public decision-making structures, which demands critical reflection on what those structures look like and their implications (Torfing, 2010; Jessop, 1998). Like governments, many civil society organizations are actively involved as agents in public policy-making, social provisioning, and discursive interventions. Second, the concept of governance in a democratic context is often grounded in the normative assumption that greater participation from non-governmental actors (e.g. the private sector, civil society organizations) is not only beneficial, but also desirable, and it is through broad-based participation that governance processes can be more effective than governments acting alone at achieving shared, public objectives (Minnery, 2007; Kooiman, 2003).

In the realm of food systems, governance spans from the internal decision-making structures of companies, non-profit organizations, and informal associations, to the more formalized policies and procedures of state governments and Indigenous authorities, and then to multilateral trade and environmental agreements. This collection focuses mainly on governance that takes place in the public realm. Issues of food governance are increasingly on the public agenda, as is evident, for example, in the development of national food policies in Canada (A Food Policy for Canada, 2018), Australia (The National Food Plan, 2013), Ireland (Food Wise 2025 Agri-Food Strategy, 2015), Scotland (National Food and Drink Policy, 2009), the United Kingdom (Food 2030 Strategy, 2010), and Wales (Food Policy and Strategy, 2010); efforts to reform the Common Agricultural Policy in the European Union (DeSchutter, 2017); and momentum to influence the United States Farm Bill and establish a national food strategy (Beyranevand & Broad Leib, 2017). While each of these efforts are grounded in conversations about the need for better integration across the arms of central governments in relation to food, they are also accompanied by conversations about the role of food systems' actors, from corporations to civil society organizations, in achieving more integrated food systems *governance* within those states.

Governance in a neoliberal context

Since the 1970s, neoliberalism has had a major impact on food systems governance across the globe (McMichael, 2009; Friedmann, 1987). Neoliberalism is a political and economic ideology rooted in the belief that well-being is best advanced by entrepreneurial freedom, limited government intervention, and the primacy of economic rationalities and the free market (Heynen et al, 2007). Unlike many other sectors, neoliberalism has not been wholly, nor uniformly, adopted within food systems. This may be best explained through the long-held view of governments that "food is different" from other commodities, as argued by agrarian theorists (Mann & Dickinson, 1978) and echoed in the title of Peter Rosset's 2006 book, *Food Is Different: Why We Must Get the WTO Out of Agriculture*.

Still, neoliberalism has had a significant impact on food governance in at least two ways. First, as described by Peck and Tickell (2002), neoliberalization has involved both a "roll back" of the Keynsian welfare state and its associated entitlements, as well as a "roll out" of new regulatory mechanisms that involve the devolution of responsibilities from the state towards public-private partnerships and voluntary compliance regimes. Roll-back neoliberalization was evident, for example, in the structural adjustment policies of the 1980s that required developing states to open up competition in previously protected agricultural commodities in exchange for further access to financial support through the World Bank and International Monetary Fund (Clapp, 2016). Roll-out neoliberalization is evident in the proliferation of voluntary fair-trade and organic standards, the rise of internal industry certification standards like GlobalGAP and increasing calls for public-private partnerships. Often, these new governance regimes rely on certification processes defined (and managed) by civil society or industry organizations. This shift towards voluntary and extra-state (often market-based) regulatory efforts is thus one of the ways that neoliberalism has had a hand in creating the conditions for the increased participation of civil society actors in food systems governance, including many associated with food movements.

These examples of roll-out neoliberalism back the points made by Kettl (2000), Ferguson and Gupta (2002) and O'Toole (1997) at the turn of the millennium that neoliberalism, and the accompanying force of devolution, was enabling new governance mechanisms through networks. These twin forces created administrative conditions that strained the capacity of local governments and increased the importance of non-governmental processes and institutions. Partnerships between governments, for-profit, and non-profit organizations developed to co-create policies and deliver state-like services (though often coming from organizations with wages substantially lower than what governments offer and with less job security), particularly at the local level. These partnerships contribute to a growing expertise of civil society actors.

The second way that neoliberalism has contributed to new governance mechanisms is through the resistance it has engendered. The critiques of neoliberalism, including the negative effects of neoliberalization in food systems, are widespread. Many see neoliberalization as involving negligible to non-existent monitoring and little public accountability (McCarthy & Prudham, 2004). In many contexts, the effects have been to undermine protections of health, equity, and sustainability (Patel & Moore, 2017). For example, in the international sphere, trade and investment agreements often serve to shelter foreign investors from domestic laws and policies, including those for environmental and labour protections (van Harten, 2017). Further, the drive for accumulation and commodification of labour and the environment, and the deepening of market logics in these areas, has made it difficult to imagine alternative social–environment relations (Guthman, 2008). However, Karl Polanyi's concept of the "double movement" reminds us that political-economic systems going through periods of market liberalization also experience social resistance (see Polanyi, 1944). While governance structures shaped by neoliberalism cause significant social and ecological challenges, they also expose cracks in the dominant system, thereby creating opportunities for increased engagement and empowerment of civil society and social movements. Social movements are seizing these opportunities to collectively advocate for a greater voice in food systems governance, especially for the most marginalized, such as food-insecure households, farm labourers, and small-scale farmers and fishers (Renting et al, 2012).

In sum, neoliberalism (and the associated process of devolution) is linked to food governance in at least two important ways. First, it has created spaces for civil society actors to engage in a variety of market-based governance efforts, including participation in the co-creation and delivery of state-like services, particularly at the local level. Second, it has contributed to the emergence of new arenas of contentious politics in which social movements have come to challenge the directions taken by governments and corporations, and demand either state re-regulation or the development of new solutions that they have a role in creating and governing. Notably, these dynamics are connected, as the roll-back of the state has led to the emergence and empowerment of a new cadre of non-government organization experts (i.e. in food security) with the administrative expertise and authority to claim a seat at governance tables on the very terms (i.e. the validation of expertise) that those tables were built upon (Martin & Andrée, 2017).

Food movements

When we use the term civil society, we are referring to the arena of social engagement that exists above the individual and below the state, in and through which individuals form their political identities (Andrée, 2007). Murphy (1994) describes civil society as the realm where "I" becomes "we". Civil society organizations (CSOs) include a multiplicity of formal and informal associations, religious organizations, and social groups – as distinct from governments and businesses – that work for what they understand to be the collective interests of society (Koc et al, 2008). Given the tensions that can emerge between different objectives in food movements, CSOs are not necessarily united in their aims. Non-governmental organizations (NGOs) are a subset of civil society organizations that are formally recognized by the states (or inter-governmental contexts) they operate in.

NGOs are usually organized as non-profit or charitable organizations and accountable to a board of citizen representatives.

CSOs of all types contribute to an array of food-oriented social movements focusing on issues from democratic accountability and environmental justice to alliances with Indigenous peoples to advance efforts of decolonization. Social movements are typically organized as informal networks that mobilize on multiple levels and in various forums around politically contentious issues. They are distinguished by an engagement in collective action with an aim to transform modes of production and consumption, as well as social organization, values, and individual well-being (Della Porta and Diani, 2006; McKenzie, 2002). While some movements work to change specific state polices or processes, their goals are generally oriented towards broader social and political change, with some movements aspiring to delegitimize government and/or transform it completely.

The food movements (and associated CSOs, broadly, and NGOs) that we focus on in this volume are made up of a wide range of individuals, groups, and organizations. These movements have no single agenda, and indeed may champion different priorities (as we discuss in Chapter 1). Still, most identify with a set of goals related broadly to creating or maintaining more sustainable, healthy, just, and democratic food systems (Constance et al, 2014). Some analysts argue that these movements find unity in what they oppose, namely the globalized, industrial, primarily market-driven food system, which mainly benefits large agri-businesses (Friedland, 2008; Allen et al, 2003). Others argue that what food movements have in common is the tendency to build practical alternatives to the globalized, industrial food system. Food movements put forward concrete, positive, and practical ideas about how better food systems can function (Roberts, 2017). As noted above, another characteristic of contemporary food movements is their increasing connectivity and sophistication, along with a tendency to adopt an integrated, systemic perspective.

While the organizations and initiatives described in each chapter emerge in specific contexts, and each may have a unique set of objectives, it is notable that all are connected with broader, often global, food movements that seek to ensure food systems are more healthy, just, and sustainable (Levkoe & Wakefield, 2013; Schiff & Levkoe, 2014; Goodman et al, 2012). Levkoe (2014) argues that "together, these collaborative efforts may be illustrative of a new wave in food activism that is represented by the emergence of a multi-scaled and cross-sectoral 'food movement' – a network of networks" (p. 399). In fact, our cases suggest one reason why food movements are increasingly connected. Food movements may start by building alternatives from the grassroots, but they eventually find they must orient themselves towards confronting and opening up established governance systems and institutionalizing, through policy reforms, an environment conducive to the alternatives they espouse. By working collectively for food systems transformation at a local level, many end up pushing for more democratic processes of engagement at various other levels of governance. And, in some cases, they find themselves pushing for, and helping to establish and implement, inclusive forms of co-governance with the state.

Theorizing food movement impacts

Several of the questions this collection seeks to address have to do with the implications of governance engagement for food movement actors themselves, and for their ability to further the aims of building more sustainable, just, and democratic food systems. These questions include: How do co-governance arrangements work in practice? What strategies of engagement are most effective in specific contexts? What are the benefits and risks of active engagement? These questions relate, in part, to the political orientations and the efficacy of the food movements. To help us answer such questions, social movement theory suggests that we need to pay attention to a combination of internal movement characteristics and the external context or "political opportunity structure" that movements engage with (Tarrow, 1998).

Shurman and Munroe (2009) explain that political opportunity structure argues that "when environmental conditions are favorable, movements are better able to mobilize and more likely to achieve their goals than when those conditions are inhospitable" (p. 158). Neoliberalism, with its associated policies and processes, is clearly an important part of this overarching context, creating obstacles and opportunities at all scales of governance. The characteristics of the immediate political and economic context that may constrain or enable movement influence include its openness to new actors, the stability or state of flux of governing political alignments, the presence or absence of possible allies within governing elites, and the state's capacity and will to repress dissent (La Forest, 2011; Schurman, 2004; Tarrow, 1998). Some analysts add a cultural element to the examination of political opportunity structure, noting that cultural context shapes people's expectations of governance institutions, their sense of their rights as citizens, and their sense of their own power (Gamson & Meyer, 1996; Nelkin & Pollak, 1981). The configuration of these structures are different at each scale: for example the dominance of large transnational agri-businesses as a result of state roll-back at the global scale, operating in relatively closed opportunity spaces, is different than the emergent political arenas being developed at the local scale, as communities and local governments grapple with the on-the-ground impacts of the global food system (Clark et al, 2015).

Internal movement characteristics are also influential on the impact of food movements (Schurman, 2004). Movement actors' level of organization, ability to gather and utilize resources, and ability to engage strategically with other actors are all relevant. Particularly important for non-profit organizations with limited funds is their ability to achieve influence by (re)framing issues in ways that capture the attention of governments and other actors (Jasanoff, 1997). Chapter 1 provides more detail on the different types of organizing frames that have been adopted by food movement actors, as well as implications of these different frames for how food movement actors engage with governance. While they may not all adopt these theoretical terms, each of the chapters in this volume pays some attention to both context and internal movement characteristics as they unpack the details of social movement engagement with food systems governance in unique contexts.

Collectively, the cases presented in this volume show that there has been a growing interest from food movements in governance processes, especially at the local level, with growing impacts regionally, nationally, and internationally. These cases align with a larger trend across the Global North. Growing engagement of food movement actors in governance is demonstrated through, for example, the emergence of a range of local or regional food policy councils (FPCs). Since 2000, FPCs have more than quadrupled in number, with over 300 across the United States and Canada in 2016 (Sussman & Bassarab, 2017). Engagement has also increased at the national level, with policy interventions led by food movements, such as the People's Food Policies in Canada (2011), Australia (2013), and the UK (2017). Each of these efforts began with social movements and non-profit organizations identifying and defining problems that could be addressed through policy initiatives (Levkoe & Sheedy, 2018; Clancy et al, 2007).

Overview of the book

Civil Society and Social Movements in Food System Governance is divided into nine chapters. Chapter 1 serves as a framing chapter for the case studies. It proposes a continuum for thinking about food movements' engagement with the state and then critically reflects on the implications of food movements playing a co-governance role. The continuum focuses on issues of participation and engagement in food systems governance and the exercise of power by social movements (from multi-stakeholderism through co-governance to polycentrism/self-governance), and a variety of opportunities and challenges to the governance options along this continuum that deserve careful reflection by academics and movement actors alike. In doing so, the chapter addresses questions such as: What forms of power do food movements mobilize to assert their position on the continuum? And, what are the benefits and limitations associated with different types of engagement in governance?

The remaining nine chapters of this volume are organized according to the main forms of governance engagement found in the cases they present. Chapters 2 through 5 shed light on the intersection between collective action for food systems transformation, mostly at local levels, and these food movement actors' engagement with governance processes. The case studies found in these four chapters are of food movement actors operating at the multi-stakeholder end of the spectrum, though these groups and movements may have aspirations for deeper governance involvement. Chapters 6 through 9 presents a series of diverse case studies that we collectively call experiments in institutionalizing co-governance.

In Chapter 2, Johnston and Andrée draw on social movement theory to examine the growing impact (but also current limitations) of two NGOs seeking to promote the integration of sustainability, health, economic development, and social justice goals within food systems governance in Canada's Northwest Territories (NWT). They focus on The Yellowknife Food Charter Coalition, a project of the Yellowknife Farmer's Market that drives action on food security issues within the NWT's capital, and Ecology North, an NGO working to influence food systems governance at the territorial level. They suggest that these NGOs exhibit several characteristics of impactful social movements, including their ability to frame challenges in novel ways, foster collaboration among a diverse set of actors, and make connections among knowledge, action, and policy. While these NGOs are becoming more than simply "stakeholders" in the NWT food system, they are not yet full and equal participants in co-governance arrangements.

In Chapter 3, MacKay and Connelly explore two co-evolving food system networks in Dunedin, New Zealand. These two networks bring to life tensions found in food movements and the trade-offs civil society groups make when engaging in governance. Our Food Network Dunedin (OFN) is a self-described grassroots organization dedicated to stimulating the production, distribution, and consumption of local food for the purpose of building a resilient and prosperous community. OFN recognized the importance of institutional resources to support food system change and were successful (along with other actors) in lobbying the city council to create a part-time position to ensure that food issues were considered across council policies. Out of these efforts emerged Good Food Dunedin, a council-led food initiative. Despite having many of the same actors involved in both networks, they are perceived differently by stakeholders involved and by the broader community, have different access to resources, and, necessarily, are taking different approaches to food system change. Reflecting on these co-evolving networks, the authors discuss how what was once a relatively small civil society effort to broaden discussion amongst existing alternative food system actors has transitioned into a broader engagement with the council to place food on the agenda. As a result, the framing of problems and solutions related to food system change has had to be reframed through negotiation and compromise. However, in this process, space has not yet emerged for critical reflection of the relative (im)mobility of concepts such as food justice, food sovereignty, food security, sustainable food, community resilience, and economic development across council-led and community-led food networks. The authors argue that while the pragmatic trade-offs and compromises might appear to be less transformative, they have resulted in dramatically extending the reach and prominence of food system change as an issue for discussion in Dunedin more broadly.

In Chapter 4, through a case study of the YYC Growers and Distributors in Calgary, Alberta, Beckie and Bacon show how local knowledge, expertise, and social capital help build a producer cooperative, which collaborates with other individuals and organizations to influence food system changes. The authors examine the development of YYC's internal, democratic self-governance model, as well as their role in the evolution of local food governance at municipal and provincial levels. As part of this analysis, they examine YYC's relationships of support and knowledge exchange between urban and rural producers; their relationships with customers and civil society organizations to raise awareness about the benefits of supporting and advancing local food systems; and their connections to municipal and provincial government representatives. Using a social economy framework, YYC serves as a link to consumers, community organizations, and government agencies, and through this pivotal role is catalyzing change in local food governance.

In Chapter 5, Levkoe and Wilson introduce the concept of prefiguration as a useful way of thinking about how collective action is related to co-governance. This chapter explores the intersections between food systems governance and social movement mobilization, examining the role of the policy-making process and the efforts of non-profit organizations and grassroots coalitions to promote empowerment, community development, and broader food systems transformation. Specifically, the authors ask how social movements can advance food policy, while also modelling alternative food futures through processes of policy development. Reflecting on a series of engagements in the Canadian government's process to develop a national food policy, they argue that by prefiguring participatory governance mechanisms and systems-based analysis, social movement networks can use policy-building processes to strengthen relationships and advance collective strategies for change.

Chapter 6 by Anderson goes another step by theorizing the impact that strong participation by citizens, civil society organizations, and social movements can have on the efficacy of governance arrangements focused on reducing hunger and food insecurity. Specifically, she is interested in the question of what happens when the public has genuine ongoing opportunities for contributing to public policy about food systems. In considering this question, she contrasts the governance mechanisms of Vermont Farm to Plate's statewide Cross-cutting Team on Food Access and the international reformed Committee on World Food Security to examine these dynamics at different scales. In reflecting on the effectiveness of each governance mechanism, characteristics that seem to play an important role are held up. These include the place of civil society and priority accorded it, especially the most vulnerable and marginalized people; the implicit or explicit rules that determine how each model functions, particularly how transparent and democratic it is; the presence of a facilitative and non-dominating "backbone organization"; use of measurable evidence-based indicators to show progress; mechanisms to allow learning and reflection; whether the work of the model structure can be integrated upwards and downwards with similar work at other scales; and whether the model facilitates coordination with more transformative social innovations such as anti-racism, wealth redistribution, and a focus on human rights.

In Chapter 7, Lowitt et al present two case studies that contrast others in the volume. While this chapter also addresses issues of co-governance, it presents a different perspective about the assumed role of the state in relation to Indigenous assertions of food sovereignty and jurisdiction over their fisheries. While the continuum presented in Chapter 1 assumes the state defines the overall contours of governance processes, Indigenous approaches to governance problematize this assumption. The chapter explores two examples of fisheries governance involving Indigenous and settler engagements, each struggling to maintain elements of self-governance and sovereignty over their fisheries in relation to the state. Collaborating with Batchewana First Nation of the Ojibways and the Saugeen Ojibway Nation, the authors seek to understand these communities' different governance aspirations, strategies used to exercise jurisdiction, and perceptions of opportunities and limitations in organizing towards enhanced equity for sustainable fisheries.

In Chapter 8, Clark explores the process of public participation design, and, more specifically, the governance of that process: Who is making the design decisions and how? The chapter's case study traces the development of a public participation design and its implementation, with keen attention to the use of power by different actors and the context within which power dynamics operated. This case is interesting because a civil society organization was on equal footing, despite power differentials, with a city health department and a county economic development and planning department in a local government planning process. Clark shows how the governance of the planning process was undergirded by an informal, trust-laden network, which enabled power-sharing and norms and values to be reconciled between organizations, resulting in an innovative public participation design. In the end, by including non-public sector actors in the process of public participation design, social equity of the planning process increased.

In Chapter 9, Clément shows the interrelationships between economic and social movement activity in rural food systems revitalization, and the central role of governance processes. This chapter explores the hybrid governance scheme developed between the state, market, and civil society to breathe life back into the small rural community of Correns, France's "first organic village". In particular, Clément looks at how Correns relied on both market-based and social approaches to local food systems governance; on the one hand, the town relied on value-adding marketing schemes through certifications including organic or Appellation d'Origine Controlée (AOC) labelling, and on the other hand, it focused on collaborative state–community initiatives. The story of Correns's transition to organic agriculture, and ultimately its adoption of a community-wide sustainable development strategy, provides an insightful example of hybrid governance in support of rural local food systems development.

Summary of the collection

Taken as a whole, the collection of cases in this volume demonstrates:

1) Food movements are seeking to have a wider, systemic impact by placing more emphasis and energy into engagement in governance processes. Over the past two decades, there have been many examples of successful local food initiatives across the Global North – from farmers' markets and urban agriculture projects to food literacy and justice for migrant worker campaigns. While celebrating these developments, the chapters in this volume point to a collective maturity of these efforts and the growing political sophistication of movement actors. Beyond focusing only on establishing innovative place-based projects, the case studies highlight collaborative efforts to have a greater impact on food systems and take seriously issues of power and social/ecological justice.
2) Social movements engaging in governance are working across a range of scales, from the local to the international. In this volume, the majority of cases speak to examples of local engagement, as this appears to be a key entry point for movements making the shift from food systems initiatives to more direct governance involvement. As expressed, there also appears to be more opportunities for communities to engage with governments at the municipal and regional level with perceived opportunities for impact. The relationship with neoliberalism is important here as there continues to be a shift in the ways governments function and the roles that non-profit organizations play in social service program delivery and as a result, in governance. As the chapters in this volume suggest, many tensions arise in social movements' relationships with government, and considering the broader context while critically reflecting on the opportunities and limits of engagement is key.
3) Social movement engagement with governance takes a variety of forms. To help interpret the strengths and weaknesses of engagement, our authors consider where their cases might be placed along a continuum, from multi-stakeholderism to polycentrism/self-governance. This enables discussion over why certain movements choose (or are limited) to engage in certain ways. It also raises important questions around the benefits and challenges associated with different levels of engagement, including the compromises that particular forms of engagement might lead to, the benefits for food systems, governments, and for other actors involved.
4) Our cases also demonstrate the central role of building relationships based on mutual trust with other actors while moving along the governance continuum. The chapters in this volume highlight the ways that many of the relationships built through local food initiatives become the foundation for broader collaborations. These relationships include farmer-to-farmer support networks, producer/harvester relationships with consumers, diverse food systems actors sitting together within

a food policy council, and Indigenous–settler relationships based on collective values and goals of food sovereignty. While there can be disagreements, actors are anchored in their desire for change. A commitment to working through tensions and being open to different ideas and approaches is essential. The work of undertaking food systems initiatives with other actors in a meaningful way can translate into deeper levels of governance engagement over the longer-term.

In conclusion, using food as a lever for social, economic, and ecological change, food movements are gaining ground by building relationships, trust, and political competency, and a result they are increasingly claiming a role in governance processes. In some cases, this is part of larger historical struggles against capitalism and colonialism, while in other cases it is about bringing the voices of the most marginalized to the fore in policy and governance. In all cases, the ongoing effects of neoliberalism, both destructive and productive, are important to take into consideration. We think these are exciting times to be working with food movement actors to chart pathways forward, and to reflect with them on the possibilities and the limitations presented by those various paths.

Notes

1 We go into more detail unpacking these different movements, and their perspectives on governance engagement, in Chapter 1.
2 That these cases are all from the Global North (or international in scope) is a product of the geographic locales in which our researchers operate, rather than a deliberate effort to focus on a particular region. We encourage similar examinations of social movement engagement in food systems governance in the Global South.

References

Allen, P., Fitzsimmons, M., Goodman, M., & Warner, K. (2003) Shifting plates in the agrifood landscape: The tectonics of alternative agrifood initiatives in California. *Journal of Rural Studies*, 19, 61–75.

Amin, S. (2011) Food sovereignty: a struggle for convergence in diversity. In: Holt-Giménez, E. (ed.) *Food movements unite! Strategies to transform our food system*. New York, Food First Books, pp. xi–xviii.

Andrée, P. (2007) *Genetically modified diplomacy: The global politics of agricultural biotechnology and the environment*. Vancouver, UBC Press.

Andrée, P., Ayres, J., Bosia, M. J., & Massicotte, M. J. (eds) (2014) *Globalization and food sovereignty: Global and local change in the new politics of food*. Toronto, University of Toronto Press.

Beyranevand, L. J. & Broad Leib, E. M. (2017) Making the case for a national food strategy in the United States. *Food & Drug Law Journal*, 72, 225.

Blay-Palmer, A., Knezevic, I., Andrée, P., Ballamingie, P., Landman, K., Mount, P., Nelson, C., Nelson, E., Stahlbrand, L., Stroink, M., & Skinner, K. (2016) Future food system research priorities: a sustainable food systems perspective from Ontario,

Canada. *Journal of Agriculture, Food Systems, and Community Development*, 3(4), 227–234. Available at: https://doi.org/10.5304/jafscd.2013.034.029

Clancy, K., Hammer, J., & Lippoldt, D. (2007) Food policy councils: past, present, and future. In: Hinrichs, C. C. & Lyson, T. A. (eds) *Remaking the North American food system*. Lincoln, University of Nebraska, pp. 227–234.

Clapp, J. (2016) *Food*. Cambridge, Polity Press.

Clark, J. K., Sharp, J. S., & Dugan, K. L. (2015) The agrifood system policy agenda and research domain. *Journal of Rural Studies*, 42, 112–122.

Clapp, J. (2016) *Food*. Cambridge, Polity Press.

Constance, D. H., Friedland, W. H., Renard, M-C., & Rivera-Ferre, M. G. (2014) The discourse on alternative agrifood movements. In: Constance, D. H., Renard, M.-C., & Rivera-Ferre, M.G. (eds) *Alternative agrifood movements: ppatterns of convergence and divergence*. Bingley, Emerald Group Publishing Limited, pp. 3–46.

Della Porta, D. & Diani, D. (2006) *Social movements: an introduction*. Second Edition. Oxford, Blackwell Publishing.

DeSchutter, O. (2017) A food policy for Europe. *Green European Journal*. Available at: www.greeneuropeanjournal.eu/a-food-policy-for-europe/

Desmarais, A., Claeys, P., & Trauger, A. (eds) (2017) *Public policies for food sovereignty: Social movements and the state*. New York, Routledge.

Ferguson, J. & Gupta, A. (2002) Spatializing states: toward an ethnography of neoliberal governmentality. *American Ethnologist*, 29(4), 981–1002.

Friedland, W. H. (2008) "Chasms" in agrifood systems: rethinking how we can contribute. *Agriculture and Human Values*, 25(2), 197–201.

Friedland, W. H. (2010) New ways of working and organization: alternative agrifood movements and agrifood researchers. *Rural Sociology*, 75, 601–627.

Friedmann, H. (1987) International regimes of food and agriculture since 1870. In: Shanin, T. (ed.) *Peasants and peasant societies*. Oxford, Basil Blackwell, pp. 258–276.

Gamson, William A. & Meyer, D. S. (1996) Framing political opportunity. In: McAdam, D., McCarthy, J. D., & Zald, M. (eds) *Comparative perspectives on social movements*. New York, Cambridge University Press, pp. 275–290.

Goodman, D., DuPuis, E. M., & Goodman, M. (2012) *Alternative food networks: knowledge practice and politics*. New York, Routledge.

Guthman, J. (2008) Bringing good food to others: investigating the subjects of alternative food practice. *Cultural Geographies*, 15(4), 431–447.

Hendrickson, M. & Heffernan, W. (2002) Opening spaces through relocalization: locating potential resistance in the weaknesses of the global food system. *Sociolgia Ruralis*, 2(4), 347–369.

Heynen, N., McCarthy, J., Prudham, S., & Robbins, P. (eds) (2007) *Neoliberal environments: false promises and unnatural consequences*. New York, Routledge.

Howard, P. H. (2016) *Concentration and power in the food system: who controls what we eat?* New York, Bloomsbury Publishing.

IPES (International Panel of Experts on Sustainable Food Systems) (2015) *IPES-Food and the new science of sustainable food systems: who shapes food systems, and who has a say in how they are reformed?* Available at: www.ipes-food.org/images/Reports/IPES_report01_1505_web_br_pages.pdf

Jasanoff, S. (1997) NGOs and the environment: from knowledge to action. *Third World Quarterly*, 18(3) 579–594.

Jessop, B. (1998) The rise of governance and the risks of failure: the case of economic development. *International Social Science Journal*, 50(155), 29–45.

Jessop, B. (2002) Liberalism, neoliberalism, and urban governance: a state-theoretical perspective. *Antipode*, 34, 452–472.

Kennedy, A. & Liljeblad, J. (eds) (2016) *Food systems governance: challenges for justice, equality and human rights*. New York, Routledge.

Kettl, D. F. (2000) The transformation of governance: globalization, devolution, and the role of government. *Public Administration Review*, 60(6), 488–497.

Koc, M., MacRae, R., Desjardins, E., & Roberts, W. (2008) Getting civil about food: the interactions between civil society and the state to advance sustainable food systems in Canada. *Journal of Hunger and Environmental Nutrition*, 3(2/3), 122–144.

Kooiman, J. (2003) *Governing as governance*. Thousand Oaks, Sage.

La Forest, R. (2011) *Voluntary sector organizations and the state: building new relations*. Vancouver, UBC Press.

Levkoe, C. Z. (2014) The food movement in Canada: a social movement network perspective. *Journal of Peasant Studies*, 41(3), 385–403.

Levkoe, C. Z. (2015) Strategies for forging and sustaining social movement networks: a case study of provincial food networking organizations in Canada. *Geoforum*, 58, 174–183.

Levkoe, C. Z. & Sheedy, A. (2018) A people-centred approach to food policy making: lessons from Canada's people's food policy project. *Journal of Hunger and Environmental Nutrition*. Doi: 10.1080/19320248.2017.1407724.

Levkoe, C. Z. & Wakefield, S. (2013) The community food centre: creating space for a just, sustainable, and healthy food system. *Journal of Agriculture, Food Systems, and Community Development*, 2(1), 249–268.

Levkoe C. Z., Koc, M., Duncan, J., Wyman, M., & Bryn, A. (2011) *People's food policy*, Discussion Paper 10 – Food Democracy and Governance. Available at: www.peoplesfoodpolicy.ca

Martin, S. & Andrée, P. (2017) Putting food sovereignty to work: civil society governmentalities and Canada's people's food policy project (2008–2011). *Journal of Civil Society*, 13(4), 374–391.

Mann, S. A. & Dickinson, J. M. (1978) Obstacles to the development of a capitalist agriculture. *The Journal of Peasant Studies*, 5(4), 466–481.

McCarthy, J. & Prudham, S. (2004) Neoliberal nature and the nature of neoliberalism. *Geoforum*, 35(3), 275285.

McKenzie, J. (2002) *Environmental politics in Canada*. Don Mills, Oxford University Press.

McMichael, P. (2009) A food regime genealogy. *The Journal of Peasant Studies*, 36(1), 139–169.

Minnery, J. (2007) Stars and their supporting cast: state, market and community as actors in urban governance. *Urban Policy and Research*, 25(3), 325–345.

Murphy, C. (1994) *International organization and industrial change*. Cambridge, Cambridge University Press.

Nelkin, D. & Pollak, M. (1981) *The atom besieged: extraparliamentary dissent in France and Germany*. Cambridge, MIT Press.

O'Toole, L. J. (1997) Treating networks seriously: practical and research-based agendas in public administration. *Public Administration Review*, 57(1), 45–52.

Patel, R. & Moore, J. W. (2017) *A history of the world in seven cheap things: a guide to capitalism, nature, and the future of the planet*. Oakland, University of California Press.

Peck, J. & Tickell, A. (2002) Neoliberalizing space. In: Brenner, N. & Theodore, N. (eds) *Spaces of neoliberalism: urban restructuring in western Europe and North America*. Oxford, Blackwell, pp. 33–57.

PFP (People's Food Policy Project) (2011) *Discussion paper 1: indigenous food sovereignty*. Montreal, Food Secure Canada.

Polanyi, K. (1944) *The great transformation*. New York, Farrar & Rinehart.

Renting, H., Schermer, M., & Rossi, A. (2012) Building food democracy: exploring civic food networks and newly emerging forms of food citizenship. *International Journal of Sociology of Agriculture and Food*, 19(3), 289–307.

Roberts, W. (2017) Radical food resistance. *Alternatives*, 43(2). Available at: www.alternativesjournal.ca/policy-and-politics/radical-food-resistance

Rosset, P. (2006). *Food is different: why the WTO should get out of agriculture*. Black Point, Zed Books.

Schiff, R. & Levkoe, C. Z. (2014). From disparate action to collective mobilization: collective action frames and the Canadian food movement. In: Leonard, L. & Kedzior, S. B. (eds) *Occupy the earth: global environmental movements*. Bingley, Emerald Group Publishing Limited, pp. 225–253.

Schurman, R. (2004) Fighting "Frankenfoods": industry opportunity structures and the efficacy of the anti-biotech movement in Western Europe. *Social Problems*, 51(2), 243–268.

Shurman, R. & Munroe, W. (2009) Targeting capital: a cultural economy approach to understanding the efficacy of two anti-genetic engineering movements. *American Journal of Sociology*, 115(1), 155–202.

Sussman, L. & Bassarab, K. (2017) *Food policy council report 2016*. Baltimore, Johns Hopkins Center for a Livable Future.

Tarrow, S. G. (1998) *Power in movement: social movements, collective action, and politics*. New York, Cambridge University Press.

Torfing, J. (2010) Local forms of governance in Denmark: the revenge of the supplement. *Journal of Power*, 3(3), 405–425.

van Harten, G. (2017) *Is it time to redesign or terminate investor-state arbitration?* Waterloo, Centre for International Governance Innovation.

Wittman, H., Desmarais, A. A., & Wiebe, N. (eds) (2011) *Food sovereignty in Canada: creating just and sustainable food systems*. Halifax, Fernwood Publishers.

1 The governance engagement continuum

Food movement mobilization and the execution of power through governance arrangements

Peter Andrée, Jill K. Clark, Charles Z. Levkoe, Kristen Lowitt, and Carla Johnston

Introduction

To enable the transition towards more sustainable and just food systems, food movements are claiming new roles in governance (Martorell & Andrée, 2018; Carlson & Chappell, 2015; Renting et al, 2012). Among academics and policy-makers, there is also growing interest in the potential contributions of social movement and civil society actors to co-governance (aka collaborative governance) processes. This interest is based, in part, on the recognition that, in social democracies, "wicked problems" (Candel, 2014) of a systemic nature can only be addressed through the active participation of all sectors: the public sector, the private sector, and civil society (Emerson et al, 2011; Blomgren Bingham et al, 2005; O'Toole, 1997). This chapter proposes a governance engagement continuum as a theoretical and practical framework to help conceptualize the role of food movement actors in food systems governance.

The governance engagement continuum helps conceptualize, compare, and critically analyze the opportunities and challenges to food movement engagement with governance across different contexts and scales. The continuum ranges from multi-stakeholderism on one end to polycentric governance (which we understand to include "self-governance") on the other. Between these poles are a range of possibilities, including some arrangements that can be defined as co-governance. Positioning the forms of social movement actor participation in governance along a continuum brings to the fore the variety of ways that civil society and social movements wrestle with power, from influencing, to sharing, claiming, and exerting power within their own contexts and within broader social, economic, and ecological systems.

This chapter begins with a discussion of the various ideological and strategic orientations of food movements found in the Global North, since these orientations have significant implications for the ways in which food movement actors engage with governance. We then discuss conceptions of power, bringing together a framework developed by Clapp and Fuchs

(2009) – which incorporates the concepts of instrumental, discursive, and structural power – with Gaventa's (2005) "power cube". The power cube distinguishes among three forms of power (visible, hidden, and invisible), and also draws attention to the diverse spaces and places where power relations are at work. Together, we believe these frames offer a useful account of what power is and how it works in relation to governance processes. Power relations of all forms are involved in the creation and implementation of governance mechanisms. They are also at work in resistance within and against governance, as well as through the claiming of new spaces outside formal governance regimes. The chapter then presents our governance continuum in detail, demonstrating its heuristic value. We end by identifying questions the continuum raises for future scholarship on food movement engagement in governance. Throughout the chapter, we draw on examples from case studies presented in this volume to illustrate our arguments.

Food movements in governance processes

In our introductory chapter, we emphasized those aspects that unite food movements. We noted a wide range of labels associated with these movements, including fair trade, civic agriculture, food justice, food sovereignty, food democracy, agroecology, slow food, and community food security (Schiff & Levkoe, 2014; Friedland, 2010). Nonetheless, we emphasized what these movements have in common: a critique of the industrial food system (Friedland 2008; Allen, 1999); a desire to build healthier, more just and sustainable food systems; practical experience in building positive alternatives (Roberts, 2013); a growing level of networking across movements (Levkoe, 2015); and the increased use of systems-thinking to diagnose and try to solve the challenges associated with how food is produced, distributed, and accessed. Here, we begin by focusing on the differences among the ideological and strategic orientations of food movements found in the Global North, since these orientations have implications for why, and how, food movement actors engage in governance processes.

Within the food systems literature, one common way to distinguish food movements, and the food initiatives they champion, is according to the strategies they employ to make change. These strategies can be understood, in part, as the orientation that movement initiatives take towards the dominant food system (McInnes & Mount, 2017). While some efforts are designed to serve as alternatives and work in parallel to the dominant food system, others directly challenge and aim to transform those systems. Still others work to achieve incremental changes within the dominant system (Holt-Giménez & Shattuck, 2011; Allen et al, 2003). Examples of the first type are often called "alternative food networks" (AFNs). In this volume, we have several case studies that focus on the construction and governance of AFNs, including the story (described in Chapter 4) of the development of a local food producer's cooperative, YYC Growers and Distributors, in Calgary, Canada.

The category of AFNs includes farmers' markets, organic supply chains, and fair-trade networks. What these diverse AFNs have in common is that they are established as alternatives to conventional supply chains by their turn away from standardized and industrial systems of food provisioning towards an emphasis on particular types of quality, locality, or production practices (Andrée et al, 2010; Goodman, 2003, 2004). The "resistance" of AFNs to the dominant, neoliberal food system is mainly enacted through their "autonomy" rather than any overt struggle or covert sabotage (Van der Ploeg, 2007, p. 3). As Van der Ploeg (2007, p. 3) writes,

> Resistance resides in the fields, in the ways in which "good manure" is made, "noble cows" are bred, "beautiful farms" are constructed, and "fresh milk" is delivered. As ancient and irrelevant as such practices may seem when considered in isolation, in the current context they are increasingly vehicles through which resistance is expressed and organized.

A great deal of ink has been spilled on the question of whether AFNs actually impact the status quo and to what extent they might reinforce problematic aspects of the dominant industrial food system (Holt-Gimenez & Shattuck, 2011; Levkoe, 2011; Fridell, 2007; Ilbery & Maye, 2005; Sonnino & Marsden, 2005; Watts et al, 2005; Guthman, 2003; Winter, 2003). The point of the distinction being drawn here is that movements focusing primarily on establishing and growing AFNs put most of their energy into building those alternatives, which has been described as slowly transitioning towards a more ideal food system through the provision of alternatives (McInnes & Mount, 2017). This focus can be contrasted with other food movements that actively seek to transform the dominant food system in one way or another. The latter, which include movements for food justice and food sovereignty, seek to actively change how the dominant system works and is structured. The nature of the changes these movements seek, however, represents another significant distinction. We return to this point below.

From the perspective of governance, the work of building a successful AFN involves considerable governance engagement on multiple levels. For example, establishing YYC Growers and Distributors in Calgary first involved building a formal internal governance structure among the producers. Developing supportive and mutually beneficial relationships with YYC's consumers, some of whom pick up community supported agriculture (CSA) shares from one another's houses, has expanded this governance structure, albeit informally. YYC also works alongside a range of community organizations, and with local food champions in municipal and provincial government, to advocate for changes that better support a more sustainable, equitable, and localized food system. Some YYC members participated in the development of Calgary's food strategy and have also provided input to the provincial government's Local Food Act. This dynamic, multi-level context that YYC is part of reveals the complexity of AFNs and local food governance issues.

Drawing on the insights of Karl Polanyi (1944) about the social reaction that regularly follows an over-emphasis on market liberalization (as we again find in the neoliberal era), AFN development is frequently presented in this literature in terms of "re-embedding" markets in social and ecological values (e.g. Vivero Pol, 2015; Pinkerton & Davis, 2015; Renting et al, 2012; Holt-Gimenez, 2011; Guthman, 2007; Sonnino, 2007). The idea of "re-embedding" markets refers to the re-regulation of food through a mix of formal (e.g. policies and laws) and informal (e.g. norms and relationships) governance mechanisms (Sonnino 2007; Andrée, 2014).

Other food movements use strategies that work from within the dominant food system, taking a reformist approach. An example in this collection is the organic movement success story of Correns' France, whereby the community has collectively engaged in global markets, but on their own terms and for their local business and broader community benefit. Their champions may be comfortable with having their initiatives embedded within a globalizing system that otherwise remains dominated by a limited number of large transnational corporations (Clapp, 2016). Even if they are uncomfortable with this situation, their work is likely to be only tangentially challenging the dominant system. This reformist label might also apply to many of the organizations that advocate for food security.

Food security means having physical and economic access to sufficient, nutritious, and culturally appropriate food (FAO, 2003). Several of the food movement efforts presented in this book, including the Yellowknife Food Charter Coalition (Chapter 2), formally organize their work under the banner of furthering food security. In his review article on the subject, Candel (2014) finds that the governance of food security is highly complex. To be effective, it requires coordination across multiple scales and sectors and the involvement of multiple actors, including international organizations, states, civil society organizations, and private sector actors. For food movement actors to call for food security is thus effectively to call for changes to food system governance. However, because of accepted food security definitions, the appeal may only be for changes to specific aspects of the food system, such as food distribution. Holt-Gimenez and Shattuck (2011) argue that "rather than call for structural change, most [food security advocates] work to increase and improve social safety nets" (p. 323). More specifically, the food security orientation, on its own, can lend itself to technocratic (rather than democratic) solutions, and is insufficiently discerning about where food comes from, the conditions under which it was produced or harvested, and the full range of impacts associated with the structures that bring food to people. Holt-Gimenez and Shattuck (2011) argue that most reformist orientations effectively reinforce the industrial food system by "fine-tuning the neoliberal project rather than encouraging substantive change of direction" (p. 123). It should be noted that movement actors have adopted approaches that have addressed some of the critiques via a community food security (CFS) approach, which focuses on long-term solutions, individual empowerment, support for local food systems, and community-based marketplaces. CFS is arguably a more alternative approach than a reformist approach (Allen, 2004).

Several points are worth making in response to the criticism of food security. First, a reformist orientation may, indeed, ultimately serve to fine-tune the neoliberal project. However, food movements are increasingly aware of this possibility, which they saw vividly in the corporate take-over of the organic food counter-culture in California in the late twentieth century (Guthman, 2004). Second, significant improvements to social safety nets – to the point where food security for all within a country is realized – would be no small political transformation, and much more than fine-tuning the neoliberal project. Perhaps most importantly, many food movement actors adopt food security as a primary goal as a way of claiming access to an ongoing, formal, governance conversation. Speaking in the language of food security (or some other reformist position) does not necessarily mean the actors in question do not have more transformative goals for their food system. In the case of the Yellowknife Food Charter Coalition (Chapter 2), efforts to establish a food charter were framed in terms of "food security", but the contents of that charter resonate more with community food security and food sovereignty (both discussed below), particularly in terms of its goals for increasing localized food production and strengthening the role of traditional (hunted, fished, and gathered) foods in the Yellowknife, Northwest Territories food system as important pathways to "food security".

In contrast to the reformist orientations of the food movements organized to promote food security, or the alterative orientations of farmers' markets, fair trade and organic farming, radical food movement orientations directly challenge the dominant industrial food system with aims of system transformation (Jarosz, 2000; Kloppenburg et al, 1996). Some consider the right to food discourse (as recognized by the Universal Declaration of Human Rights, the United Nations Covenant on Economic, Social and Cultural Rights, and the United Nations General Comment 12) as a radical call to action. For example, the Civil Society Mechanism to the Committee on World Food Security (Chapter 6) strongly supports the right to food and adequate nutrition and has staunchly resisted efforts to water that down to strictly food security. As discussed in Chapter 6, some states, such as Brazil, have formally adopted the right to food in their constitutions in response to social movement mobilization. While no state fully guarantees this right in practice, human rights discourse has a considerable degree of moral authority. Champions see it as an accountability framework for the marginalized and least powerful to use when they are interacting with powerful actors and governments (McKeon, 2017). In deliberate contrast to food being considered a commodity or private good, which limits access to food to those that can pay, promotes accumulation that only enriches a few, and creates negative environmental, social, and economic externalities (Sumner, 2011), advocates also call for food to be governed as part of a commons (Vivero Pol, 2015). This approach would recognize food as a common good to which everyone has equal access. The right to food is an aspirational goal and its implementation is being actively monitored by the Civil Society Mechanism.

For food movements with ambitions to transform the dominant food system, food sovereignty has become an increasingly important frame, galvanizing those working under the related banners of food justice, Indigenous food systems, and the right to food (Edelman et al, 2014; Schiff & Levkoe, 2014). The concept of food sovereignty was developed through collaborative dialogue among global social movement organizations in response to multiple crises within the dominant food system and its neoliberal ethos of privatization and corporatization (Wittman et al, 2010; Patel, 2009). Originally rooted in the experiences of peasant farmers and Indigenous peoples across the globe, the food sovereignty movement now includes a wide range of fishers, landless peoples, workers, migrants, pastoralists, forest communities, women, youth, eaters, and environmental and urban movements actively engaged in struggles to transform food systems (Desmarais, 2007). Food sovereignty focuses on the idea that people, especially those most marginalized by the dominant food system, are reclaiming control of their own food systems. At its core, food sovereignty asserts the principle that all people have the right to food and that democracy and gender equality are fundamental to this realization (Patel, 2009). While food sovereignty developed as a grassroots social movement, more recently many activists and advocates have turned their demands into public policy in an attempt gain visibility and power (Desmarais et al, 2017). For example, social movements have led efforts to incorporate food sovereignty into governance at the state level in Ecuador, Bolivia, and Venezuela, although with mixed impacts (Clark, 2015). In the Global North, the food sovereignty movement is growing, as evidenced by the story told in Chapter 5 of Food Secure Canada's (FSC) efforts to influence national food policy in Canada. Food sovereignty shapes how FSC and its allies organize and what they ask of their governments, though the institutional uptake from the Canadian state is unlikely in the short term to go as far as it has in other countries.

The mixed impacts achieved by radical social movements when they seek to influence formal governance institutions are to be expected. Many movement actors that engage in formal channels of policy advocacy, design, and implementation recognize that this engagement typically requires making compromises (Andrée et al, 2011; Levy & Newell, 2005). These processes are situated in institutions and narratives that have often been formed without their influence, and these actors are trying to put them to new uses. Some movement actors try to avoid compromise by building new governance spaces entirely. Indigenous food sovereignty movements, in particular, regularly call for self-determination and independence from the state for their food systems, based on the recognition that Indigenous peoples have sustainably managed food systems from time immemorial, and that capitalism and continued settler-colonialism (wherein the state is a central actor) disrupts these practices (Rudolph & McLachlan, 2013; Corntassel, 2012; Morrison, 2011). Chapter 7 presents the efforts of two First Nations to establish self-governance arrangements for their fisheries in their traditional territories as an important part of food sovereignty.

One might expect food governance engagement by food movement actors motivated by these more radical orientations to look quite different from the types of engagement undertaken by the champions of AFNs or system reformers. And indeed, many do come to governance engagement from a different angle. The first question many radical food movement actors ask is whether their goal is to have legitimacy and influence at formal decision-making tables set up by others, or whether their goal is to dismantle those tables altogether and build new ones? Some food movement actors do refuse to engage with governance mechanisms established by others, and their efforts often look a lot like that of the AFN champions who build alternatives outside of the mainstream. In fact, the social movements that initially led the organic farming movement in Britain and the United States (Guthman, 2004), as well as the fair trade movement (Fridell, 2007), started in this way. These movements were initiated by radicals who chose to go their own way and build up governance mechanisms (e.g. peer-based certification of organic production) that did not require any real engagement with the industrial food system or even the states they arose within. Chapter 5 also addresses these questions in relation to FSC's efforts to engage its members in the Canadian government's national food policy consultations.

Further to these questions, Gerlach and Hine (1970) argue that a dichotomous perspective on whether change should come from inside versus outside of a system is never productive. They recognize that change can be initiated from outside a system, but if those in power in the current system accept the method of change, then it quickly becomes an inside change. Furthermore, if those in power consider the change as "progress" then it is simply a developmental rather than a "radical" change. If those in power resist change, then those who desire it must mobilize collective action to oppose those in power. The logic of this position for flexibly working both inside and outside of formal governance institutions resonates with Stevenson et al's (2008) argument for a multifaceted approach to food system transformation through the interconnected struggles of warriors (i.e. interventionism), builders (i.e. creating alternatives), and weavers (i.e. developing strategic and conceptual linkages).

Finally, as noted in the book's Introduction, food movements are increasingly connected with one another, regionally and globally (Levkoe, 2015; Constance et al, 2014). Movements for sustainable agriculture, fair trade systems, strong social safety nets, Indigenous self-determination, and increased democratic participation (as espoused by food justice and food sovereignty) have become more networked, learning from each other's successes and challenges and working collaboratively. As food movements become more closely linked, some actors become the warriors, some the builders, and some the weavers (Stevenson et al, 2008). And all must engage with governance in one way or another. The builders of alternatives, like the YYC Growers and Distributors, must create internal governance mechanisms and engage with other actors in the ways described above. The warriors for Indigenous food sovereignty, while claiming the status of nations, must also engage with the states in which they are embedded to create effective co-governance resource management regimes (Chapter 7). Meanwhile, the weavers seeking to shape local (Chapter 8), sub-national (Chapter 6), or national policies (Chapter 5), informed by the goals of community food security or the principles of food sovereignty, must all learn to understand and appreciate the perspectives of a range of stakeholders, including those within the dominant food system as well as fellow food movement activists, in order to be able to exercise power within those governance processes.

Power and governance

As Lang and Heasman (2015) point out, governance is about the execution of power. Governance processes, whether formal decision-making structures or informal collaborations, are themselves also manifestations of power relations. Getting a strong handle on power is thus fundamental to understanding both how governance structures come to be and how governance works (McKeon, 2017). Here, we bring together a framework introduced by Clapp and Fuchs (2009), along with Gaventa's (2005) "power cube", which distinguishes between forms, places, and spaces of power. When combined, these frameworks offer multiple ways of considering how food movements experience and work with power in governance processes.

Clapp and Fuchs (2009) distinguish among three types of power as a capacity that is possessed and used by actors: instrumental, discursive, and structural power. Each form is related and interacts with the others. Instrumental power means wielding power and influence over others through direct action, fueled by the use of resources (technical, financial, informational, etc.) (McKeon, 2017; Clapp & Fuchs, 2009). Food movement actors rely on instrumental power when they strategically marshal resources to achieve their goals (McCarthy & Zald, 1977). The power states wield to ensure compliance with laws, through their right to impose fines or prison terms, is also an example of instrumental power (Fuchs et al, 2011). The private sector has used instrumental power to physically assault and kill activists (Global Witness, 2018) and to influence states to negotiated favorable trade agreements with other states, and civil society organizations employ instrumental power when they collectively, and effectively, lobby for the expansion of safety net programs (Winson, 1993).

Discursive power refers to the ability of actors to frame issues, develop dominant narratives, and establish new norms (McKeon, 2017;

Cheyns & Riisgaard, 2014; Clapp & Fuchs, 2009; Andrée, 2007). Groups of all kinds employ discursive power to frame problems in a way that matches their values and priorities, thereby setting up the logical solutions to those problems (Schön, 1993). Framing is critical in governance because it determines the way in which authority is claimed and it defines the purposes for the use of authority. For example, conceptualizing a food system as a commons versus a market results in different logics regarding who the legitimate governance parties are and how problems are best addressed. To wield discursive power, a group must be considered legitimate by its own constituency and by other political actors. This involves a level of trust in their expertise and intentions.

Structural power is the power to define the scope and institutional structures in which decisions are made. Institutions and their decision-making processes establish the range of behaviours and choices made available to actors (Clapp & Fuchs, 2009). Structural power can set agendas and legitimize actors that are to be engaged in these processes. Those who successfully utilize structural power effectively "build the table" for decision-making and are not just invited to the table. For food movements to play a greater role in governance, this type of power is often necessary to exercise so that governance processes including them move from multi-stakeholderism to some form of co-governance or self-governance. For example, Batchewana First Nation (Chapter 7) is actively seeking to build their own table for governing fisheries in their territory, rather than be invited to other existing fisheries management forums at which they will only be one voice. Further, Chapter 5 describes how FSC designed and implemented an inclusive and democratic decision-making process to demonstrate how principles and values of social movements can be integrated into decision-making structures.

Torangeau (2017) points out, building on Foucault (1978) and his interlocutors (e.g. Dean, 2010; Andrée, 2007), that there is another type of power not accounted for in Clapp and Fuchs' (2009) framework. Clapp and Fuchs (2009) focus on the conscious deployment of power by actors. This needs to be distinguished from the constitutive power of norms and discourses, which have no immediate source but rather permeate the governance environment as a result of previous power relations. More precisely, the "discursive power" of agents that Clapp and Fuchs (2009) speak of is only possible within a larger discursive "field" of social and cultural norms. These are the "power relations in which . . . forms of agency appear" (Dean, 2010, p. 461). Examples of constitutive power include the inherent legitimacy accorded to some ontological or epistemological understandings over others. In governance conflicts, constitutive power can be critically important by, for example, according legitimacy to the knowledge of scientifically trained "experts" over the traditional knowledge of Indigenous peoples, fishers, or farmers.

Gaventa's (2005, 2007) power cube (see Figure 1.1) adds further depth to our understanding of power in relation to governance. Composed of the interrelated dimensions of spaces, levels, and forms, the power cube considers

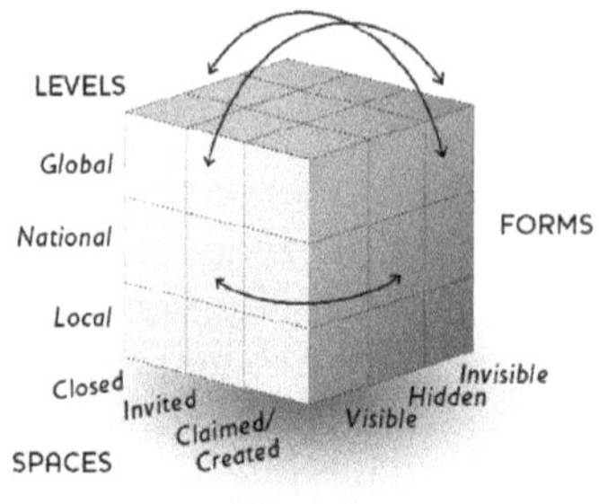

Figure 1.1 Gaventa's power cube: the levels, spaces and forms of power
Source: Gaventa, 2005

"how spaces for engagement are created, the levels of power (from local to global), as well as different forms of power across them" (Gaventa, 2005, p. 6). The power cube has been used in a range of political settings and sectors by activists and researchers to map out power dynamics and identify possible avenues for social change and transformative action.

We introduce the power cube here to add complexity to Clapp and Fuchs' (2009) framework presented above in three main ways. First, instrumental power (e.g. a state funding agricultural programs or establishing new food safety regulations) is normally a visible (formal) execution of power. However, the power cube reminds us that instrumental power may equally be informal and hidden from view, such as when a bribe changes hands, or a favour is promised, that influences a decision. Discursive power can be thought of as taking various forms according to the power cube. It may be visible to some (such as the repeated use of what many recognize as a lie), but to those who believe the lie, the discursive power may remain hidden. Meanwhile, the constitutive power of widespread social norms, such as the assumption that land or other resources actually belong to settlers in a colonial context, even though these were never ceded to them by Indigenous people (see Borrows, 2010), can best be thought of as a hidden, if not invisible, form of power.

Another important contribution of the power cube is to remind us that governance spaces are not neutral. Power relations shape who can enter certain spaces of participation and what is possible within them. The "spaces" of power on the power cube add nuance to Clapp and Fuchs' (2009) description of structural power. In the process of exercising power, food movements will encounter both closed and invited spaces, and spaces within which it is easier or harder to claim power (Gaventa, 2006). How food movements seek change (such as within or outside of the dominant system) may influence the types of spaces they encounter. For example, movement actors may find certain spaces "closed" to them (that is, decisions being made by a certain set of actors behind closed doors). Such spaces may need to be pried open, the act of which involves executing "structural" power. Invited spaces are more welcoming. They may range from one-off consultations to more ongoing processes of engagement led, for example, by a government or other agency. However, Clapp and Fuchs' idea of structural power reminds us that these spaces were initially structured by others, and movement actors need to carefully assess how effectively they can use them to further their own goals. The case study of FSC's engagement with the national food policy process in Canada (Chapter 5) offers an excellent example of the opportunities, tensions, and internal challenges that an "invited" governance space can create for food movement actors.

Sometimes, when closed and invited spaces simply do not serve their purposes, movement actors choose to claim new spaces of governance. Where governance structures are weak or absent around particular issues or scales, it may be attractive for social movement actors to create spaces for themselves. This is evident at the community level, where historically the state and transnationals have less activity, and, therefore, food governance structures tend to be emergent (Clark et al, 2015; Mendes, 2008). Chapter 8 describes how new spaces were co-created with movement actors given the lack of hardened governance structures. In Chapter 2, non-governmental organizations (NGOs) urge governments to engage with new and broader governance spaces, ones which benefit from collaborative efforts.

In a certain way, all spaces are "claimed", with closed and invited spaces often gaining legitimacy simply because they were created ("claimed") by the state and thereafter recognized by other states and political actors. Some spaces claimed by food movement actors eventually get this same formal recognition. The creation of governance and regulation for organic foods, initially organized through social movements, but eventually legitimized by the state, is an example of a movement-claimed space that eventually became a formal governmental space. The claiming of new spaces is an act of sovereignty, and today we often see this action taken most strongly by food movements associated with food sovereignty and/or Indigenous self-governance. For example, Chapter 7 shows how two First Nations claiming governance of fisheries within their traditional territories are challenging the authority of the settler-colonial state.

Finally, the power cube reminds us that governance engagement is always a multi-level (or multi-scalar) exercise, even if certain levels need more attention than others. The efforts pursued by food movements operate at a variety of scales and political-economic contexts, and this has implications for the types of governance arrangements that may be sought by movement actors. For example, in Canada the federal government and each of the provinces and territories are granted powers to govern in certain areas and to share jurisdiction in others. Food policy engagement with the state may need to consider and address municipal, provincial/territorial, federal, and transnational scales. Food movement actors may find that they can wield power more effectively at some scales versus others and sometimes "jump" between scales to impact change. In addition, Indigenous communities have a separate set of jurisdiction and governance powers over particular lands, populations, and activities related to food systems. This is based on a number of constitutionally recognized agreements between Indigenous peoples and the Crown that are the foundation for First Nations to relate to the Canadian government on a nation-to-nation basis. Despite successive governments failing to adhere to these original agreements, various court cases have reaffirmed the powers of Canada's Indigenous peoples (First Nations, Inuit, and Métis people) including section 35(1) of the Constitution Act, 1982, that affirmed Aboriginal and Treaty Rights. In sum, the power cube defines the levels as local, national, and global, but there are other levels to consider, including other layers of government, in addition to internal power relations (and governance mechanisms) among social movement actors themselves.

Drawing on this theoretical discussion of governance and power, we now turn to a description of the governance engagement continuum. While all food movement actors are seeking to change food systems in some way, the engagement continuum assists us in considering their different ambitions in terms of changing governance in food systems as well as the changes they have, or have not, been able to influence.

The governance engagement continuum

Building on existing scholarship focused on social movements and governance as well as the case studies found in this volume, we propose a continuum of ways that food movements mobilize resources to disrupt, influence, or engage in the execution of power through food systems governance arrangements (see Figure 1.2). The position of a governance arrangement on this continuum can enable or restrain the ability of food movement actors to participate in the exercise of power over decision-making in food systems.

On the far left of the continuum, multi-stakeholderism has food movement actors as simply one set of groups seeking to exercise influence in a pluralistic public sphere, with government (albeit rarely a neutral actor) usually formally holding the reins of decision-making and public policy or program execution. Food movement actors seek influence within multi-stakeholder governance processes, but ultimately remain "governed" (or repressed) by the state. Here, power from state actors typically result in control from the top. State and market actors (as enabled by the state) lead problem-setting, rule-setting, and enforcing compliance, e.g., taxing and subsidization from the government, pricing from market mechanisms. If social movements have a role in governance, it is likely in the form of a stakeholder of one or more civil society groups.

On the far right of the continuum, we have the opposite ideal type, with social movement actors participating in the active governance of their own

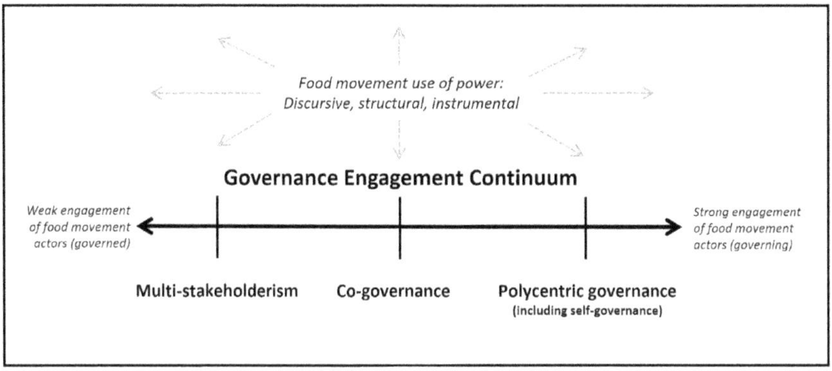

Figure 1.2 Governance engagement continuum: The role of food movements

food systems (or distinct parts of those systems) through self-governance. Here, social movements exert bottom-up control, framing issues and designing institutions and compliance structures that re-embed social and environmental concerns in the food system to address those issues. However, even in cases of self-governance, social movement activity always takes place under the jurisdiction of one or more states (even if the actors involved contest the state's sovereignty). It also requires coordination with other state and market systems (Carlisle & Gruby, 2017; Ostrom, 2009). These two points leads us to position self-governance under polycentrism, which assumes the co-existence of multiple centers of power and control.

Between these two poles is a range of emerging arrangements that involve food movement actors sharing governance roles with other actors, including state agencies. This middle zone is the area that we are especially interested in exploring through the case studies in this volume, given the growth of experiments across the Global North taking place in what is variously termed "regulatory pluralism" (MacRae & Winfield, 2016) and "collaborative governance" (Emerson & Nabatchi, 2015). Moving from left to right, we find governance structures in which food movement actors increasingly wield the defining instrumental, structural, and discursive power over their food systems.

Governance is not only about the formal exercise of power through institutions. Civil society organizations are known to be adept at wielding discursive power within the broader public sphere in order to shape public debate (Dryzek, 2013; Wapner, 1995). For example, they are known to use their instrumental power to influence broader governance outcomes outside of their formal participation as a stakeholder in a government consultation or as a participant, for example, on a food policy council. These informal uses of power are recognized in Figure 1.2 by noting, across the top of the continuum, a food movement's use of instrumental, discursive, and structural power within and outside of governance arrangements.

Where a specific governance arrangement is found on this continuum is a result of a combination of at least three important factors, which can be related back to the discussion of the execution of power. The first two factors speak to the agency of these social movement actors, while the third speaks to structure or context.

The first factor is the sum total of the instrumental resources of the food movement actors and their allies, such as financial and technical capacity, as well as actor's ability to mobilize these resources successfully. Broadly speaking, the more resources at their disposal, the more likely they can participate in a co-governance arrangement with other actors, though that depends also on whether such an arrangement is what they want (which is one aspect of the second factor, below), and what other actors in the system they can accept (an aspect of the third).

The second factor is social movement actors' own strategic or ideological orientation, which refers to their understanding of food systems goals and how they want to achieve those ends (from alternative to reformist to transformative). This factor speaks to their willingness to engage in, claim, or accept, one form of governance or another. In some cases, a specific orientation (e.g. food security) may lead an actor to call for minimal changes to the governance structures of a food system, while those holding more a transformative orientation (e.g. food sovereignty) may work to build governance mechanisms that share power more equitably among particular actors. However, it is important not to assume that actors holding more transformative orientations necessarily achieve governance arrangements towards the right side of the continuum (Cadieux & Slocum, 2015). To link back to our discussion of power, it is also notable that an actor's strategic and ideological orientation, such as the right to food, may initially be wielded as a form of discursive power and be highly visible as such. However, it can eventually become constitutive in a Foucauldian sense (and "hidden", to use power cube terminology), though both of these results depend on whether other relevant actors come to accept the logic of this discursive frame.

The political-economic context is the third key factor that helps to define the position of a governance initiative on the continuum. Social movement actors seek a certain level of control within the food systems, but this means either wrestling control from others, whether voluntarily or through political maneuvering (the successful execution of all three forms of power), or by creating space for more collaborative forms of control. In general terms, the political-economic context can be thought of as a confluence of economic, institutional, and discursive forces that define the field of political engagement (Andrée, 2007; Gill & Law, 1989; Gramsci, 1971). It refers to existing and entrenched relations of power. In some instances, this context may readily accommodate food movement participation in governance processes on terms these actors seek, but many other decision-making fora are clearly captured by specific private interests or deeply beholden to norms and values that simply do not mesh with the aspirations of the movement actors (and thus are either visibly or invisibly "closed"). As noted above, these situations sometimes lead movement actors to create ("claim") new governance spaces. If social movements are not able to achieve the degree of shared power in a governance arrangement that they seek, this is most likely explained by the way their aspirations are hampered by the confluence of political-economic forces at that historical moment.

At the national and global level, private actors wield extensive power in food governance arrangements, in what some would call a "corporatocratic" governance arrangement (Suarez-Villa, 2012). Clapp (2016, p. 24) sums up the power relations of the dominant food system well: "In the dominant globalizing food system, power tends to be concentrated in the private sector with transnational corporations, and to a lesser extent with the states that create the regulatory spaces within which these transnationals operate". This reality effectively means many of the formal governance spaces at these levels are "closed" to real participation by food movement actors.

Given the importance of these private sector actors in food governance, especially at the international level, one of our chapter authors suggested that we actually add an additional axis to Figure 1.2 of the governance continuum, highlighting weak to strong private sector engagement in these same governance processes. That would have brought additional insights forward, such as can you have strong food movement and strong private sector involvement at the same time? What happens when one of these actors is strongly involved, and the other only weakly engaged?

While these are important questions to explore, we chose not to make our continuum more inclusive. Instead, we decided to simply have our continuum take for granted that in the dominant, market-based food system, governance always includes all three sectors – public, private, and civil society – to a greater or lesser extent. This certainly applies in the Global North, as these three sectors exist in any community and at all scales.

We also note that Indigenous peoples are not identified separately in the continuum, although one of our chapters (Chapter 7) deals with co-governance between two First Nations and the Canadian state, and it uses the continuum to situate their struggle for co-governance and self-governance. We recognize that Indigenous peoples are not social movements, though they have been at the forefront of various food movements around the world. Indigenous peoples have a distinct governance relationship with the state. Many effectively claim nationhood, though this claim of sovereignty is complicated by the fact that their territories are embedded within internationally recognized sovereign states, and they must continually negotiate with those states as they assert their claims.

To conclude this discussion of where a governance arrangement lies on the continuum with reference to our discussion of power, the more that food movement actors are able to define governance arrangements in the ways they want (whether to the left or right on the continuum), the more they have been able to execute structural power. We now turn to describing each of the positions on our governance continuum in more detail.

Multi-stakeholderism

Multi-stakeholder governance describes social movement actors working as "stakeholders", holding a singular stake in food systems, directly with or against the state and/or corporations. Central to this type of neoliberal governance arrangement is the desire to engage a wide range of stakeholders that have their own interests and then have these stakeholders compete and bargain to integrate their interests in decision-making in a zero-sum game (Cheyns & Riisgaard, 2014). Process dynamics tend to be defined by conflict and competition, which is heavily influenced by interest group politics and pluralism (Clapp & Fuchs, 2009). Here, stakeholders tend to use various forms of power to influence outcomes, not collaborate on outcomes.

Multi-stakeholder decision-making processes are largely defined by the state and corporations, with emphasis given to market-oriented processes. In the neoliberal context, corporate actors, particularly those that work transnationally, are more in the position to exert structural power in determining the rules. McKeon (2017, p. 6) highlights that corporate power in the global food regime makes the current form of multi-stakeholderism in governance structures a systemic problem:

> corporations participate in multi-stakeholder platforms and public-private partnerships (PPPs) only where the framework is in line with their business strategies. Conversely, PPPs tend to align their standards to what corporations want in order to obtain their adhesion, so they are unlikely to act as a force for transformation of corporate practices.

Multi-stakeholder governance often provides food movements a consultative role, but little more. There is no promise to integrate interests or collaborate on solutions (Ansell & Gash, 2008). Civil society stakes and interests are considered to be on the same footing as state and market actors, ignoring power imbalances (McKeon, 2017). The limitations FSC faced in trying to create a more inclusive state-led food policy process (Chapter 5) is an example of this. Considering civil society as a stakeholder can reinforce power differentials (Cheyns & Riisgaard, 2014). At best, food movement actors can make change within the system if they have the power to achieve outcomes that match their objectives. This is noted in Chapter 6 about the promising work of the Committee on World Food Security and the Vermont Farm to Plate Network. At worst, being designated a stakeholder can be used to legitimize decisions that do not incorporate food movements' interests, and participation by these actors can blunt the ability to challenge corporate or state actors (Lang & Heasman, 2015).

Co-governance

Co-governance can be defined as multiple actors working together to meet shared governance goals, versus managing competing interests (Kooiman, 2003). But these arrangements are about more than simple coordination. Co-governance refers to a middle ground between the hierarchical nature of most governmental processes and the self-organization of social movement or market processes (Somerville & Haines, 2008). Co-governance is not new, but its use as a policy instrument is increasing as it is recognized that non-state actors have resources and capacities needed to solve complex public problems (Ansell & Gash, 2008; Emerson & Nabatchi, 2015). Co-governance often means that collaborators have shared objectives for food system change. In this way, Sumner (2011) and Vivero Pol (2015) see co-governance structures as the way to realize the commoning of food. Chapters 3 and 8 illustrate co-governance arrangements in cities in New Zealand and the United States as local governments and civil society collaborate to achieve common objectives.

Blomgren Bingham et al (2005, p. 551) document a variety of emerging forms of co-governance in practice, with the connection between them a

> mechanism for cooperation and coordination among diverse and often rival participants in the policy process. As a result, these processes may increase the likelihood of a stable agreement and may contribute positively to participants' sense of justice, fairness, and the perceived legitimacy of the institutions involved.

As such, co-governance schemes are typically designed to be flexible, reflexive and adaptive so that social learning can take place to design the best solutions (Voß et al, 2006).

Collaborative governance, as a type of co-governance arrangement, is the processes and structures that engage people across sectors to "carry out a public purpose that could not otherwise be accomplished" (Emerson & Nabatchi, 2015, p 18). Collaborative governance is also focused on problem solving and consensus versus managing competing stakes (Ansell & Gash, 2008). Institutional procedural norms and rules are used to foster collaboration (as opposed to ground rules for guiding behaviour) (Emerson & Nabatchi, 2015). Trust, shared understanding, commitment to process, and "thick" communication are cornerstones of collaborative governance. Further, power and resource asymmetries are recognized and a commitment is made to create processes that empower disadvantaged collaborators at the outset (Ansell & Gash, 2008). In collaborative governance arrangements, civil society participants are co-producers of governance outcomes, and seek equal stewardship of the process (Paquet & Wilson, 2011). Participants can exert discursive and structural power by co-problem setting and co-developing arrangements, such as process rules for decision-making (Ansell & Gash, 2008). Social movements may go as far as being the conveners of a collaborative governance arrangement, inviting stakeholders from other sectors to the table.

Polycentric governance, including self-governance

First defined by Elinor Ostrom in the early 1970s, polycentric governance refers to organizational structures in which a number of independent actors coordinate their relationships with one another under an agreed-upon set of rules (Araral & Hartley, 2013). When brought to the realm of public policy, polycentrism does not simply imply decentralization. Rather, it implies first the recognition (among all actors involved) of existing self-governing processes (Nagendra & Ostrom, 2012). While we considered placing self-governance as a separate category on the far right of our continuum, we ultimately decided to include self-governance within the realm of polycentrism based on the recognition that self-governance processes are necessarily situated within nested and often overlapping scales of governance and jurisdiction (Carlisle & Gruby, 2017). Nation-to-Nation relationships between First Nations and the state are a good example of this arrangement in its ideal. Chapter 7 provides an example of two First Nations striving for a form of polycentric governance and Chapter 5 demonstrates FSC's ideal of food sovereignty as another example of this.

Ostrom's life work involves specifying the conditions under which polycentric systems emerge, so that public policy could "develop institutions that bring out the best in humans" (Ostrom 2010, p. 25) While never suggesting polycentrism as a panacea for addressing all challenges, her research emphasized the value of exploring diverse polycentric institutions because of their potential to help "innovativeness, learning, adapting, trustworthiness, levels of cooperation of participants", and the ability to achieve "more effective, equitable and sustainable outcomes at multiple scales" (Ostrom 2010, p. 25).

However, it should be noted that there is little evidence that polycentric governance has been widely accepted by formal institutions. Too often, formal institutions tend not to recognize the existing (often informal) self-governing processes of local communities or customary authorities (Rahman et al, 2017). Darnhofer and Strauss (2015, pp. 38–39; see also Gibson-Graham, 2008) write in a report from the Rethink Farm Modernisation and Rural Resilience project:

> while the potential for the establishment of polycentric systems is present, it seems that the (informal) networks and their potential contribution to regional resilience is not perceived by formal institutions and policy makers which seem to be heavily committed to a clearly defined idea of modernization . . . Similarly, the established institutions and policy makers have limited appreciation for non-capitalist practices . . . Farmers and many stakeholders however . . . frame farming as

> more than a submission to "the bottom line" or the "imperatives of capital" and welcome economic diversity.

This quote speaks to the fact that the constitutive frame within which governance negotiations take place is so important because frames can legitimize, or in this case, delegitimize informal governance efforts.

Reflections on food movements engagement in governance

The governance engagement continuum is not intended to be a judgement on governance arrangements by suggesting that one arrangement is more or less desirable. Rather, the aims of the social movements within their strategic and ideological orientations, in many ways, indicate the types of governance arrangements sought by groups. In the example provided earlier, some orientations such as food security may be more reformist in nature, and, therefore, may only seek to wield power to influence an outcome within the dominant neoliberal food system. In this case, effectively influencing decision-making within a group as a stakeholder may be desirable. In contrast, an orientation of food sovereignty, which is more transformational in aspiration, may lead to efforts to exert a deeper role in governance and may reject co-governance with certain other actors (e.g. multinational corporations). Having said that, social movements are, by definition, about change. The more power exerted by groups or coalition of groups associated with food movements, and their ability to govern decision-making, the greater the ability to transform the system. And in the past few decades, as food movements gain authority/influence, more co-governance experiments are emerging.

It is important to note that the continuum is an analytical tool, a simple heuristic for a complex system that is filled with various tensions. We recognize that presenting power as three distinct types, notwithstanding the nuances added by the power cube, is a bit simplistic since in reality the exercise of power is not separated out like this in the day to day. There is also a complex net of regulations/power relations that shape any governance system. Simultaneous processes can occur where multiple systems are interfacing with multiple actors and multiple forms of power are exercised. Our continuum really needs to be nested within another level of governance, and it in turn needs to be nested in other levels of governance. Historical political relations that provide the context within which power is exercised also cannot be ignored, and, in this way, food movements employing power cannot simply engage in governance in one space, disengage in a second, and then claim power in a third. Lastly, food movements are not monolithic. The chapters in this collection illustrate the diversity of ways the food movement is engaging in governance, reflecting various points along this continuum.

References

Allen, P. (1999) Reweaving the food security safety net: mediating entitlement and entrepreneurship. *Agriculture and Human Values*, 16(2), 117–129.

Allen, P. (2004) *Together at the table: sustainability and sustenance in the American agrifood system*. University Park, Penn State Press.

Allen, P., FitzSimmons, M., Goodman, M., & Warner, K. D. (2003) *Alternative food initiatives in California: local efforts address systemic issues*. Santa Cruz, UC Santa Cruz, Center for Agroecology and Sustainable Food Systems.

Andrée, P. (2007) *Genetically modified diplomacy: the global politics of agricultural biotechnology and the environment*. Vancouver, University British Colombia Press.

Andrée, P. (2014) The neoliberalizing environmentalities of alternative agri-food networks in Australia: a critical examination of challenges and possibilities. In: Andrée, P., Ayres, J., Bosia, M., & Massicotte, M.-J. (eds) *Globalization and food sovereignty: global and local change in the new politics of food*. Toronto, University of Toronto Press, pp. 141–172.

Andrée, P., Dibden, J., Higgins, V., & Cocklin, C. (2010) Competitive productivism and Australia's emerging "alternative" agri-food networks: producing for farmers' markets in Victoria and beyond. *Australian Geographer*, 41(3), 307–322.

Andrée, P., Cobb, M., Moussa, L., & Norgang, E. (2011) Building unlikely alliances around food sovereignty in Canada. *Studies in Political Economy*, 88, 133–162.

Ansell, C. & Gash, A. (2008) Collaborative governance in theory and practice. *Journal of Public Administration Research and Theory*, 18(4), 543–571.

Araral, E. & Hartley, K. (2013) *Polycentric governance for a new environmental regime: theoretical frontiers in policy reform and public administration*. Paper presented at international conference on public policy, Grenoble, France, 26–28 June.

Blomgren Bingham, L., Nabatchi, T., & O'Leary, R. (2005) The new governance: practices and processes for stakeholder and citizen participation in the work of government. *Public Administration Review*, 65(5), 547–558.

Borrows, J. (2010) *Drawing out law: a spirit's guide*. Toronto, University of Toronto Press.

Cadieux, K. V. & Slocum, R. (2015) What does it mean to do food justice? *Journal of Political Ecology*, 22(1). Doi: http://dx.doi.org/10.2458/v22i1.21076.

Candel, J. J. L. (2014) Food security governance: a systematic literature review. *Food Security*, 6(4), 585–601.

Carlisle, K. & Gruby, R. (2017) Polycentric systems of governance: a theoretical model for the commons: polycentric systems of governance in the commons. *Policy Studies Journal*. Doi:10.1111/psj.12212.

Carlson, J. & Chappell, M. J. (2015) *Deepening food democracy*. Minneapolis, Institute of Agriculture and Trade Policy.

Cheyns, E. & Riisgaard, L. (2014) The exercise of power through multi-stakeholder initiatives for sustainable agriculture and its inclusion and exclusion outcomes. *Agriculture and Human Values*, 31(3), 409–423.

Clark, P. (2015) Can the state foster food sovereignty? Insights from the case of Ecuador. *Journal of Agrarian Change*, 16(2), 183–205.

Clark, J. K., Sharp, J. S., & Dugan, K. L. (2015) The agrifood system policy agenda and research domain. *Journal of Rural Studies*, 42, 112–122.

Clapp, J. (2016) *Food*. Cambridge, Polity Press.

Clapp, J. & Fuchs, D. A. (eds) (2009) *Corporate power in global agrifood governance*. Cambridge, MIT Press.

Constance, D. H., Hendrickson, M., Howard, P. H., & Heffernan, W. D. (2014) Economic concentration in the agrifood system: impacts on rural communities and emerging responses. In: Bailey, C. C., Jensen, L. & Ransom, E. (eds) *Rural America in a changing world*. Morgantown, West Virginia University Press, pp. 16–35.

Corntassel, J. (2012). Re-envisioning resurgence: indigenous pathways to decolonization and sustainable self-determination. *Decolonization: Indigeneity, Education & Society*, 1(1), 86–101.

Darnhofer, I. & Strauss, A. (2015) *Organic farming and resilience (Austria) Case Study Report D3.3*. RETHINK research project. Available at: www.wiso.boku.ac.at/fileadmin/data/H03000/H73000/H73300/PJ/rethink/WP3_CaseStudyReport_AT.pdf

Dean, M. (2010) *Governmentality: power and rule in modern society*. Los Angeles, Sage.

Desmarais, A. A. (2007) *La Vía Campesina: globalization and the power of peasants*. Halifax, Fernwood.

Desmarais, A. A., Claeys, P., & Trauger, A. (eds) (2017) *Public policies for food sovereignty: social movements and the state*. New York, Routledge.

Dryzek, J. (2013) *The politics of the Earth: environmental discourses*. New York, Oxford University Press.

Edelman, M., Weis, T., Baviskar, A., Borras Jr, S. M., Holt-Giménez, E., Kandiyoti, D., & Wolford, W. (2014) Introduction: critical perspectives on food sovereignty. *Journal of Peasant Studies*, 41(6), 911–931.

Emerson, K. & Nabatchi, T. (2015) *Collaborative governance regimes*. Washington, Georgetown University Press.

Emerson, K., Nabatchi, T., & Balogh, S. (2011) An integrative framework for collaborative governance. *Journal of Public Administration Research*, 22 (1), 1–29.

FAO (Food and Agriculature Organization of the United Nations) (2003) *Trade reforms and food security*. Rome, FAO.

Foucault, M. (1978) *The history of sexuality: Volume 1*. New York, Random House.

Fridell, G. (2007) *Fair trade coffee: the prospects and pitfalls of market-driven social justice*. Toronto, University of Toronto Press.

Friedland, W. H. (2008) "Chasms" in agrifood systems: rethinking how we can contribute. *Agriculture and Human Values*, 25(2), 197–201.

Friedland, W. H. (2010) New ways of working and organization: alternative agrifood movements and agrifood researchers. *Rural Sociology*, 75(4), 601–627.

Fuchs, D., Kalfagianni, A., & Havinga, T. (2011) Actors in private food governance: the legitimacy of retail standards and multistakeholder initiatives with civil society participation. *Agriculture and Human Values*, 28(3), 353–367.

Gaventa, J. (2005) *Reflections on uses of the "power cube" approach for analyzing the spaces, places, and dynamics of civil society participation and engagement*. Brighton, Institute of Development Studies.

Gaventa, J. (2006) Finding the spaces for change: a power analysis. *IDS Bulletin*, 37(6), 23–33.

Gaventa, J. (2007) *Levels, spaces and forms of power: analysing opportunities for change' in power in world politics*. In: Berenskoetter, F. & Williams, M. J. (eds) *Power in world politics*. New York, Routledge, pp. 204–224.

Gerlach, L. P. & Hine, V. H. (1970) *People, power, change: movements of social transformation*. Indianapolis, MacMillan.

Gibson-Graham, J. K. (2008). Diverse economies: Performative practices for other worlds. *Progress in Human Geography*, 32(5), 613–632.

Gill, S. & Law, D. (1989) Global hegemony and the structural power of capital. *International Studies Quarterly*, 33(4), 475–499.

Global Witness (2018) *At what cost? Irresponsible business and the murder of land and environmental defenders in 2017*. Available at: www.globalwitness.org/fr/campaigns/environmental-activists/at-what-cost/

Goodman, D. (2003) The quality "turn" and alternative food practices: reflections and agenda. *Journal of Rural Studies*, 19, 1–7.

Goodman, D. (2004) Rural Europe redux? Reflections on alternative agro-food networks and paradigm change. *Sociologia Ruralis*, 44(1), 3–16.

Gramsci, A. (1971) *Selections from the prison notebooks*. New York, International Publishers.

Guthman, J. (2003) Fast food/organic food: reflexive tastes and the making of "yuppie chow". *Social & Cultural Geography*, 4(1), 45–58.

Guthman, J. (2004) *Agrarian dreams: the paradox of organic farming in California*. Los Angeles, University of California Press.

Guthman, J. (2007) The Polanyian way? Voluntary food labels as neoliberal governance. *Antipode*, 39(3), 456–478.

Holt-Gimenez, E. (2011) Food security, food justice, or food sovereignty? Crises, food movements, and regime change. In: Alkon, A. & Agyeman, J. (eds) *Cultivating food justice: race, class and sustainability*. Cambridge, MIT Press, pp. 309–330.

Holt Gimenez, E. & Shattuck, A. (2011) Food crises, food regimes and food movements: rumblings of reform or tides of transformation? *Journal of Peasant Studies*, 38(1), 109.

Ilbery, B. & Maye, D. (2005) Alternative (shorter) food supply chains and specialist livestock products in the Scottish–English borders. *Environment and Planning A*, 37(5), 823–844.

Jarosz, L. (2000) Understanding agri-food networks as social relations. *Agriculture and Human Values*, 17(3), 279–283.

Kloppenburg, J., Hendrickson, J., & Stevenson, G. W. (1996) Coming in to the foodshed. *Agriculture and Human Values*, 13(3), 33–42.

Kooiman, J. (2003) *Governing as governance*. Thousand Oaks, Sage.

Lang, T. & Heasman, M. (2015) *Food wars: the global battle for mouths, minds and markets*. London, Routledge.

Levkoe, C. Z. (2011) Towards a transformative food politics. *Local Environment*, 16(7), 687–705.

Levkoe, C. Z. (2015) Strategies for forging and sustaining social movement networks: a case study of provincial food networking organizations in Canada. *Geoforum*, 58, 174–183.

Levy, D. & Newell, P. (2005) *The business of international environmental governance*. Boston, MIT Press.

MacRae, R. & Winfield, M. (2016) A little regulatory pluralism with your counter-hegemonic advocacy? Blending analytical frames to construct joined-up food policy in Canada. *Canadian Food Studies/La Revue canadienne des études sur l'alimentation*, 3(1), 140–194.

Martorell, H. & Andrée, P. (2018) The commoning of food governance in Canada: lessons for a national food policy. In: Vivero Pol, J. L., Ferrando, T., de Schutter O., & Mattei, U. (eds) *Routledge handbook of food as a commons*. New York, Routledge.

McCarthy, J. & Zald, M. (1977) Resource mobilization and social movements: a partial theory. *American Journal of Sociology*, 82(6), 1212–1241.

McInnes, A. & Mount, P. (2017) Actualizing sustainable food systems. In: Koc, M., Sumner, J., & Winson, A. (eds) *Critical perspectives in food studies*. Don Mills, Oxford University Press, pp. 332–347.

McKeon, N. (2017) Are equity and sustainability a likely outcome when foxes and chickens share the same coop? Critiquing the concept of multistakeholder governance of food security. *Globalizations*, 14(3), 379–398. Doi: 10.1080/14747731.2017.1286168.

Mendes, W. (2008) Implementing social and environmental policies in cities: the case of food policy in Vancouver, Canada. *International Journal of Urban and Regional Research*, 32(4), 942–967.

Morrison, D. (2011) Indigenous food sovereignty: a model for social learning. In: Wittman, H., Desmarais, A. A., & Wiebe, N. (eds) *Food sovereignty in Canada: creating just and sustainable food systems*. Halifax, Fernwood, pp. 97–113.

Nagendra, H. & Ostrom, E. (2012) Polycentric governance of multifunctional forested landscapes. *International Journal of the Commons*, 6(2), 104–133.

Ostrom, E. (2009) Polycentric systems as one approach to solving collective-action problems. In: Salih, M. (ed.) *Climate change and sustainable development*. Cheltenham, Edward Elgar.

Ostrom, E. (2010) Beyond markets and states: polycentric governance of complex economic systems. *The American Economic Review*, 10(3), 641–672.

O'Toole, L. J. (1997) Treating networks seriously: practical and research-based agendas in public administration. *Public Administration Review*, 57(1), 45–52.

Paquet, G. & Wilson, C. (2011) Collaborative co-governance as inquiring systems. *Optimum Online*, 41(2), www.optimumonline.ca/browse.phtml?cat_id=4.

Patel, R. (2009) Food sovereignty. *The Journal of Peasant Studies*, 36(3), 663–706.

Pinkerton, E. & Davis, R. (2015) Neoliberalism and the politics of enclosure in North American small-scale fisheries. *Marine Policy*, 61, 303–312.

Polanyi, K. (1944) *The great transformation: the political and economic origins of our time*. New York, Farrar & Rinehart Inc.

Rahman, H. M. T., St. Ville, A., Song, A., Po, J., Berthet, E., Brammer, J., Brunet, N., Jayaprakash, L., Lowitt, K., Rastogi, A., Reed, G., & Hickey, G. (2017) A framework for analyzing institutional gaps in natural resource governance. *International Journal of the Commons*, 11(2), 823–853.

Renting, H., Schermer, M., & Rossi, A. (2012) Building food democracy: exploring civic food networks and newly emerging forms of food citizenship. *International Journal of Sociology of Agriculture and Food*, 19(3), 289–307.

Roberts, W. (2013) *The no-nonsense guide to world food*. Second edition. Toronto, New Internationalist.

Rudolph, K. R. & McLachlan, S. M. (2013) Seeking Indigenous food sovereignty: origins of and responses to the food crisis in northern Manitoba, Canada. *Local Environment*, 18(9), 1079–1098.

Schiff, R. & Levkoe C. Z. (2014) From disparate action to collective mobilization: collective action frames and the canadian food movement. In: Leonard, L. &

Kedzior, S. B. (eds) *Occupy the earth: global environmental movements. Advances in Sustainability and Environmental Justice*, Special Issue, 15, 225–253.

Schön, D. A. (1993) Generative metaphor: a perspective on problem-setting in social policy. In: Ortony, A. (ed.) *Metaphor and thought*. Cambridge, Cambridge University Press, pp. 137–163.

Somerville, P. & Haines, N. (2008) Prospects for local co-governance. *Local Government Studies*, 34(1), 61–79.

Sonnino, R. (2007) Embeddedness in action: saffron and the making of the local in southern Tuscany. *Agriculture and Human Values*, 24(1), 61–74.

Sonnino, R. & Marsden, T. (2005) Beyond the divide: rethinking relationships between alternative and conventional food networks in Europe. *Journal of Economic Geography*, 6(2), 181–199.

Stevenson, G. W., Ruhf, K., Lezberg, S., & Clancy, K. (2008) Warrior, builder, and weaver work. In: Hinrichs, C. & Lyson, T. (eds) *Remaking the American food system: strategies for sustainability*. Lincoln, University of Nebraska Press, pp. 33–62.

Suarez-Villa, L. (2012) *Globalization and technocapitalism: The political economy of corporate power and technological domination*. Burlington, Ashgate Publishing Company.

Sumner, J. (2011) Serving social justice: the role of the commons in sustainable food systems. *Studies in Social Science*, 5(1), 63–75.

Torangeau, W. (2017) GMO doublespeak: an analysis of power and discourse in Canadian debates over agricultural biotechnology. *Canadian Food Studies*, 4(1), 108–138.

Van der Ploeg, J. D. (2007) *Resistance of the third kind*. Paper presented to the ESRS conference, 23 August 2007, Wageningen. Available at: www.jandouwevanderploeg.com/EN/publications/articles/resistance-of-the-third-kind/

Vivero Pol, J. L. (2015) Transition towards a food commons regime: re-commoning food to crowd-feed the world. January 13. Available at: https://ssrn.com/abstract=2548928. Doi: http://dx.doi.org/10.2139/ssrn.2548928

Voß, J. P., Bauknecht, D., & Kemp, R. (2006) *Reflexive governance for sustainable development*. Cheltenham, Edward Elgar Publishing

Wapner, P. (1995) Politics beyond the state: environmental activism and world civic politics. *World Politics*, 47(3), 311–340.

Watts, D. C. H., Ilbery, B., & Maye, D. (2005) Making re-connections in agro-food geography: alternative systems of food provision. *Progress in Human Geography*, 29(1), 22–40.

Winson, A. (1993) *The intimate commodity: food and the development of the agro-industrial complex in Canada*. Toronto, Garamond Press.

Winter, M. (2003) Embeddedness, the new food economy and defensive localism. *Journal of Rural Studies*, 19(1), 23–32.

Wittman, H., Desmarais, A. A., & Wiebe, N. (eds) (2010) *Food sovereignty: reconnecting food, nature and community*. Halifax, Fernwood.

2 Pathways to co-governance?

The role of NGOs in food governance in the Northwest Territories, Canada

Carla Johnston and Peter Andrée

Introduction

Food systems in Canada's Northwest Territories (NWT) offer a distinct context for examining food movement advocacy, compared to agriculture-based food systems. These systems are strongly rooted in fishing, hunting, and gathering, but are also heavily dependent on imported foods. The agricultural sector is only nascent, but burgeoning, and there are extensive interactions between Indigenous and settler food systems.[1] Further, the NWT has particularly high rates of food insecurity and diet-related disease, and growing difficulties in accessing traditional land-based foods. When it comes to governance, the NWT is both heavily and lightly governed. It is heavily governed by many layers of jurisdictional interest in the North, from municipal, territorial, and federal governments, to several Comprehensive Land Claim Agreements (between the federal and territorial governments and Indigenous peoples) that define Indigenous self-government and co-management of resources in parts of NWT. It is lightly governed by virtue of the fact that only 40,000 people live on a territory ten times the size of England, and that devolution of governmental responsibilities of the NWT from federal to the territorial elected assembly only started in 1967, with the final devolution occurring in 2013. As a result of the above factors, governance bodies responsible for managing resources and people, including the territories' food systems, have limited capacity and are "light" on the ground. Together, this collection of factors means that the NWT (see Figure 2.1) represents a distinctive food environment for social movement actors seeking to further the goals of sustainability, health, economic development, and justice.

This chapter examines the emerging role of two non-government organizations (NGOs) in food governance processes: the Yellowknife Food Charter Coalition (the "Coalition"), which works on food security issues within the city of Yellowknife; and Ecology North, an NGO working on food systems issues at the territorial level. As part of their work, both organizations seek to influence food governance issues, at their different scales. The primary author of this chapter has worked directly with both NGOs in the context of a Participatory Action Research (PAR) project since 2016. Our goal is to unpack the elements that allow for NGOs to have an impact in food system governance and to assess what this means for the potential for co-governance. Drawing on social movement theory and collaborative governance theory, we show how the Coalition and Ecology North exhibit several characteristics of impactful social movements, as described in this volume's introduction, including their ability to frame challenges in novel ways, to foster collaboration among a diverse set of actors including the state, and to make the connections between knowledge, action, and policy. We also examine the political opportunity structure they are working in. Our work identifies those aspects of the structure that enable social movement engagement, as well as some that constrain them.

Chapter 1 in this book proposes a governance engagement continuum (from multi-stakeholderism to co-governance) as a way of understanding social movement participation in food system governance. The analysis presented below positions Ecology North and the Coalition's involvement in NWT and Yellowknife food policy, respectively, as stakeholders on left side of this continuum. However, we show that these NGOs are having a growing influence on efforts made by governments in Yellowknife and the NWT to develop new food-related policies, presenting the potential to move to the right on the continuum towards co-governance.

Theorizing NGO impact in governance

Chapter 1 argues that power can be exercised in at least three ways by actors involved in governance processes: instrumentally, structurally, and discursively. These distinctions are helpful for understanding the types of power that social movement actors work with, and within, but they offer little explanation of the conditions that allow a social movement's exercise of power in relation to governance processes to be effective. Social movement theory fills this gap, as discussed in the Introduction to this volume. Social movement theory's explanation of impact pays close attention to both internal movement characteristics (e.g. movement structure, organization, strategies, ability to mobilize resources, and ability to frame issues (Schurman, 2004)) and the external context, or "political opportunity structure" (e.g. the context's openness to new actors, or the stability of governing alignments) that movements engage with (Tarrow, 1998; Schurman, 2004).

The literature on NGOs within the environmental movement adds further insight to internal movement dynamics, emphasizing the relationship between issue framing, movement organization, and strategy. For example, Jasanoff (1997) argues that environmental NGOs have a growing role in governance because of their ability to build the "knowledge–action link" in three specific ways. First, NGOs tend to include "technical experts critical of accepted frameworks of environmental knowledge and regulatory policy" (p. 583) who know how to grab media attention and mobilize public opinion around alternative policy frames. Second, NGOs are often able to build more inclusive policy communities (or "epistemic networks") around defined policy objectives than the state can. Since policy success can be dependent on ordinary citizens accepting "the validity of official policy framings and to participate in their implementation" (p. 581), NGOs' abilities to bridge lay-expert, activist-professional, and local-global divides, as they conceptualize and advance policy prescriptions, become important resources for everyone involved in governance. Finally, a key dimension of NGOs' ability to build the knowledge-action link is to have "boots on the ground", so to speak. NGO activities in technology transfer, information gathering, policy implementation, monitoring, and enforcement all contribute to the growing effectiveness of NGOs to influence, and participate in governance processes. In the context of a discussion of co-governance – which typically involves some combination of actors from civil society, the private sector, and the state – Schurman and Munro (2009) advise we also pay attention to the way that economic factors such as industry characteristics, the organization of production and consumption, and the influence of economic institutions shape a social movement's overall opportunity structure.

Whereas social movement theory provides a framework for understanding the conditions that allow movement actors to have an impact on governance processes, the co-governance literature helps us understand what this particular form of governance looks like in practice. As noted in Chapter 1, drawing on Kooiman (2003), co-governance is most simply defined as multiple actors working together to meet shared governance goals. There is a growing literature on various forms of collaborative governance involving some combination of private sector, civil society, and state actors (see Paquet & Wilson, 2011; Voss et al, 2006; Kooiman, 2003). Based on a synthesis of existing frameworks and practice-based knowledge, Emerson et al (2011, p. 2) offer a narrower definition of collaborative governance which we employ here. For them, co-governance refers to

> the processes and structures of public policy decision making and management that engage people constructively across the boundaries of public agencies, levels of government, and/or the public, private and civic spheres in order to carry out a public purpose that could not otherwise be accomplished.

This definition is narrower than that of Kooiman because it presupposes a *voluntary* collaborative regime, situated in what is traditionally considered the realm of *public* policy. These delineations hold for our case study of these NGOs, but Emerson et al's (2011) definition may not be appropriate for other contexts where collaborative governance is emerging (e.g. in the context of the exercise of treaty rights by First Nations, as discussed by Levkoe and Lowitt in this volume).

Emerson et al (2011) provide a series of ten propositions to further describe the dynamics that enable and sustain a collaborative governance regime. These ten propositions can be categorized into four broad elements of capacity for joint action. First, one or more central drivers for collaboration must be present. The main drivers are leadership, consequential incentives, interdependence, and uncertainty. Second, "quality principled engagement" among actors must be generated. This refers to working across boundaries to reach a common goal through open communication amongst representatives from all relevant interests. Such engagement is sustained by "interactive processes of discovery, definition, deliberation, and determination". Third, a "shared motivation" must be fostered through "repeated, quality interactions that foster trust, mutual understanding, internal legitimacy, and shared commitment" (Emerson et al, 2011, p. 11). Finally, this "virtuous cycle" of principled engagement and shared motivation should contribute towards sustainability, in terms of the "institutional arrangements, leadership, knowledge, and resources" that generate sustained capacity for joint action. Compared to the broad definition of co-governance offered by Kooiman (2003), Emerson et al's (2011) sets a high bar with the depth of commitment and interaction that they find in functional co-governance arrangements. To be effective and sustained over time, collaboration must be needed as well as valued, and it must be practiced in ways that build mutual understanding, commitment, and capacity for joint action among the collaborators in the governance regime. These principles, gleaned from effective co-governance arrangements, offer a bar against which to measure other situations, like that found in the NWT, where social movement organizations and NGOs are actively seeking to establish a co-governance relationship with the state and other actors.

Our focus in this chapter is on two NGOs active in Yellowknife and NWT food systems, how they shape municipal and territorial governance processes anchored by the state, and whether those processes enable civil society collaboration in governance (i.e. co-governance). Together, social movement theory alongside Emerson et al's contributions to co-governance theory, allow us to hone in on the internal and external factors influencing potential collaborative processes as we address these questions.

Methodology: PAR in the North

This chapter is grounded in PAR undertaken by the lead author in partnership with the Coalition, Ecology North, and FLEdGE (Food: Locally Embedded, Globally Engaged) research group.[2] PAR is a critical research methodology focused on creating partnerships that work together to change the status quo through informed action. "Participatory" refers to collaborative processes whereby participants identify problems together and co-develop research methods and outputs to begin to address a local problem. "Action" signifies that this is not just research for its own sake. PAR has the explicit goal of mobilizing the results of the research process through action (Kindon et al, 2007). Finally, "Research" signifies ongoing observation, reflection, and analysis of the problem and research results by all involved in the process (Chevalier & Buckles, 2013).

To put this methodology into practice, the primary author worked over two summers (2016 and 2017) to support the Coalition and Ecology North in their food systems initiatives, with a focus on political advocacy and policy work. The primary research question that guided this work was created in collaboration with the partners. It asks: How can civil society groups, local businesses, community members, and decision-makers from the City of Yellowknife engage around the vision and principles of the Yellowknife Food Charter to improve the policy arena for a just and sustainable food system for Yellowknife?

To answer this question, the research included participant observation by working directly with these organizations on their initiatives and actions. Then, these experiences were related back to literature on food systems and resource management in the North as well as policy documents from Yellowknife and other Canadian cities, such a municipal food strategies and community plans. This chapter draws heavily on these "insider" experiences and documents as the main sources of evidence. This chapter tells the story, from the primary author's perspective, of working with Ecology North and the Coalition to strengthen food systems governance.

Ecology North is a territorial NGO based in Yellowknife, founded in 1971. Ecology North's mission is "Bringing people and knowledge together for a healthy northern environment" (Ecology North, 2017a). Through this mission, Ecology North encourages public participation in resolving environmental issues through a commitment to environmental, social, and community well-being. Its program areas focus on climate change, waste reduction, water quality, local food production, and environmental education (Ecology North, 2017a).

The Yellowknife Food Charter Coalition is an informal civil society organization created in 2014 through the Yellowknife Farmers Market. The Coalition is made up of local businesses and organizations, food producers, harvesters and foragers, nutritionists and other health practitioners, members of the Yellowknives Dene First Nation (YKDFN), and other interested individuals. As the Coalition states,

> A Food Charter sets a community mandate for broad-based action on local food security. . . [It] promotes a just and sustainable food system for all Yellowknifers . . . [and provides] a point of entry for groups and individuals to gather, generate ideas, and identify how to collectively respond and create projects that increase food security for Yellowknifers.
> (YKFM, 2016)

Seeing the need to create space for integrated solution-building, the Coalition created a food charter as a guide for diverse actors within the Yellowknife food system. The Coalition is mainly focused on municipal, Yellowknife-based actions and policy; however, they recognize that territorial and federal contexts impact their city setting. Both Ecology North and the Coalition work within the context of food systems in the NWT and Yellowknife.

Food system context in the NWT and Yellowknife

Hunting, fishing, gathering, agriculture, and imported foods all have a role to play in NWT food systems. In terms of hunting, fishing, and gathering, they have been a form of food provisioning by the Indigenous peoples in the NWT since time immemorial. There has also been a long history of state interference in traditional hunting and fishing practices in the NWT. The first federal act for the preservation of game was enacted in 1894 (Dene Nation, 2016). This and subsequent Acts took a paternalistic approach that saw the Indigenous peoples of the NWT as incapable of conserving wildlife (Sandlos, 2007), largely ignoring their thriving land-based ways of life. Fishing, hunting, and gathering continue to be a strong part of Indigenous food systems in the NWT and they are practiced by many non-Indigenous people as well.

While agricultural production is currently minimal, enthusiasm and activity for local food production, including farming, is on the rise. In terms of the history of farming in the NWT, many of the missions and fur trading posts in the NWT in the late 19th century reported acreage for subsistence. During the 20th century, farming efforts were supported by the federal government through sub-stations of the Dominion Experimental Farm system in Fort Smith, Fort Resolution, and Fort Providence by 1911, as well as Fort Good Hope in 1928 and Fort Simpson in 1947 (Leahey, 1954). These farms have had a lasting impression in the communities they were a part of. Many, including Indigenous growers, look back to these farms as inspiration for northern agriculture today. At the same time, these farms are part of the legacy of colonialism as they were privileged by patriarchal power structures that sought to expand the "modernized" economy into the North through agriculture (Sandlos, 2007). Further, gardening and farming were included in residential schools[3] in the NWT that Indigenous children were forced to attend. There have been other agricultural enterprises and projects throughout the NWT over time, however, the sector has remained very small.

Currently, imported foods make up a large portion of food consumed in the North. Deepening ties to imported foods in the NWT largely occurred with the boom of resource extraction in the post-World War Two years. In another iteration of modernizing efforts, the federal government strongly supported the resource extraction industry though subsidies for railways, highways, airports, and hydroelectric developments around key mining areas (Sandlos, 2015). This infrastructure made the importing of foods into the NWT much easier. At the same time, industrial food products were getting their own subsidies, both hidden and explicit, that made them seem economically efficient and could undercut the food grown in the NWT (Friedmann, 1992).

There are real challenges associated with NWT food systems. Food insecurity in the NWT is the second highest in Canada at the provincial/territorial level, with over 24 per cent of households experiencing moderate to high levels (Tarasuk et al, 2016; Council of Canadian Academies, 2014). The obesity rate in the NWT is 10 per cent higher than the Canadian average (GNWT HSS, 2011). The nutrition transition among Indigenous populations in the NWT is well documented, with a shift from land-based diets to imported foods of low nutritional value with detrimental health effects (Sheehy et al, 2014; Zotor et al, 2012; Sharma et al, 2009; Receveur et al, 1997). Climate change affects all parts of the food system, and particularly hunting, fishing, and foraging as these practices rely on the integrity and continuity of local ecosystems (Spring, 2018; Wesche et al, 2016; Wesche & Chan, 2010).

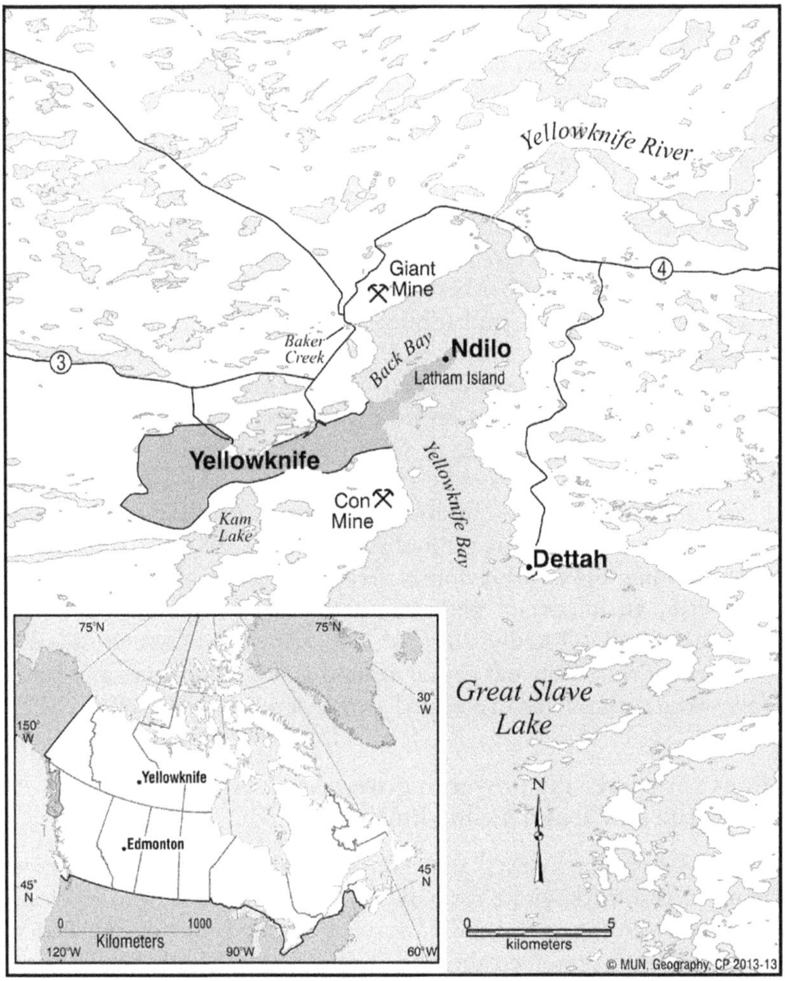

Figure 2.1 Yellowknife, Giant Mine, and the Dene settlements of Ndilo and Dettah. Map by Charlie Conway

Source: Sandlos & Keeling, 2016 (used with permission)

As the capital of the NWT, the Yellowknife food system has much in common with the rest of the territory, but also has several distinct qualities. Yellowknife is the largest urban centre in the NWT, containing around half of the population. It is a multi-cultural city of Indigenous peoples and settlers from within Canada and around the world. The Yellowknife region is part of the traditional lands of the YKDFN,[4] which now has permanent settlements in Ndilo (located with the municipal borders of Yellowknife) and Dettah see (Figure 2.1). There are other Indigenous peoples living in Yellowknife, including those of the North Slave Metis Alliance[5] and members of other First Nations. Yellowknife also has the largest population of non-Indigenous people in the NWT. While food insecurity rates are not specified for the city, the *Yellowknife, Ndilo and Dettah Food Assessment* (Lutra Associates Ltd, 2010) notes high use of emergency food services driven by the increasing price of nutritious food and high cost of housing, putting stress on people vulnerable to food insecurity. Urban gardening is prevalent in Yellowknife, with over 300 community gardeners (YKCGC, 2017) and many home gardens. Commercialized urban agriculture is minimal, but there is growing interest and initiatives through the Yellowknife Farmers Market and others. Rural food-producing areas near the city and in other regions of the NWT contribute to the local food supply for the city. Traditional foods are an active part of the city's food system with 37 per cent of Indigenous households hunting or fishing and 20 per cent gather berries (NWT Bureau of Statistics, 2014a, 2014b). Hunting, harvesting, and fishing is also widely practiced by non-Indigenous people in the city. However, the environmental legacy of arsenic pollution from Giant Mine, just outside of Yellowknife (see Figure 2.1), has had a significant negative impact on local gathering and fishing practices. As part of processing the gold ore at Giant, arsenic trioxide was released untreated through the mine's smoke stack, falling on the land and lakes around the mine. As well, 237,000 tonnes of arsenic trioxide are currently stored in underground chambers (O'Reilly, 2015). The YKDFN, "described the area of the mine site as a previously very productive valley full of blueberries and fish, an important gathering area that the mining operation completely destroyed" (O'Reilly, 2015, p. 334). Currently, there are advisories not to swim, drink, fish, and gather around many of the lakes and lands in the Yellowknife region due to elevated arsenic levels (GNWT HSS, 2017).

Conditions for exercising power in governance: Growing NGO influence in the NWT and Yellowknife

This section analyzes the internal and external conditions that the Coalition and Ecology North work within for building more sustainable and just food systems in Yellowknife and the NWT, drawing on social movement theory. In line with the theory presented above, first we show how both the Coalition and Ecology North exhibit several of the characteristics of impactful social movement actors, including their ability to frame challenges in novel ways, and to make the connections between knowledge, action, and policy. We then examine the political opportunity structure they are working in, arguing that the political and economic context of NWT, and in particular what we have defined as the fact that it is both lightly and heavily governed, has important consequences for the potential for social movement engagement in governance. This analysis contributes to our subsequent assessment of where these NGOs are situated on the governance continuum presented in Chapter 1 of this book.

Impactful social movements: Internal characteristics

One of the key internal characteristics of effective NGOs is their ability to frame problems and solutions in new ways that then "legitimate and motivate collective action" (McAdam et al, 1996). NGOs are often critical of status quo knowledge and policy, and they also play a constructive role mobilizing public opinion and policy communities, around alternative issue frames (Jasanoff, 1997). Ecology North and the Coalition are framing food issues in a new way in Yellowknife and the NWT. Both call for a comprehensive approach to food systems governance that considers hunting, gathering, fishing, agriculture, and imported foods from a food systems lens. They also hope to reshape governance, seeking to build a stronger food system through collaboration among civil society actors and governments. While these framings draw on the shared knowledge of other food movements, they are a novel approach in territorial and municipal jurisdictions of the NWT and Yellowknife, respectively.

At the municipal scale, the creation of the Yellowknife Food Charter itself offers a strong example of how the actors involved in the Coalition bring a new perspective to the issue of food insecurity in Yellowknife. The Charter frames just and sustainable food systems in Yellowknife in a way that challenges traditional silos and encourages collaborative action. The vision and principles of the Charter have subsequently been mobilized by the Coalition and its allies to put pressure on local government to address food policy. Specifically, during the 2015 elections at the municipal, territorial, and federal levels, the Coalition used the Charter to draw media attention, and to talk with politicians, businesses, civil society organizations, and members of the public about the importance of food policy (YKFM, 2016). Coalition members then used the Charter as the basis of presentations to Yellowknife City Council about the need for a city-wide food strategy (City of Yellowknife, 2015, 2016a, 2017). These efforts succeeded in putting food policy on the radar of the City of Yellowknife. In the City of Yellowknife *Council's Goals and Objectives 2016–2018*, developing an agriculture strategy (albeit not a comprehensive "food strategy" – a point we return to below) was included for the first time (City of Yellowknife, 2016b). Further, as part of the City's

2018 Citizens Budget Survey, the agriculture strategy received a majority "yes" vote. It was one of the only budget items to do so, showing the legitimacy the Coalition's vision had gained in the eyes of Yellowknife residents (City of Yellowknife, 2018a).

Territorially, Ecology North is also pushing forward an alternative vision for reshaping NWT food systems. They envision,

> [the GNWT creating] a cross-sector Territorial Food Strategy . . . where fishing, hunting, gathering, agriculture and imported foods would all be considered together, along with their impacts on health, employment and economic development, the traditional economy and the environment. This would mean bringing together the various sectors and departments within the GNWT that focus on different parts of the food system to collaborate on the food system as a whole . . . [They also envision] a territorial-wide food organization that provides a platform for advocacy and community voices about NWT food systems. This organization would have formalized connections to decision-makers within the GNWT to provide policy suggestions for the Territorial Food Strategy and provide discussion and feedback on existing policies and actions with the food system.
>
> (Ecology North, 2017b)

In line with this vision, Ecology North, the Yellowknife Farmers Market, and the Northern Farm Training Institute penned a letter to the minister of the GNWT Department of Industry, Tourism and Investment about the creation of a Territorial Food Strategy (Ecology North, 2017b), however this letter has not received a response. As well, Ecology North has moved to create the NWT Food Network (as described below).

The impact of both NGOs needs to be understood in the context of the fact that they do more than try to influence policy (Schurman, 2004; Jasanoff, 1997). They do work on policy and governance issues, but they also work on the ground. As a result, the NGOs are able to tailor their message to policy-makers through constant feedback from other stakeholders and the public, and they are able to reach citizens directly to inform their views of the issues at hand. Further, they actively communicate their vision to research institutions, funding agencies, and others, not just to governments. This ability to work at multiple levels to influence discourse is one of the ways that NGOs are most powerful (Jasanoff, 1997; Wapner, 1995). It allows them to find synergies between their instrumental power (e.g. how they spend their limited time and money in the community) and their discursive power (e.g. how they frame issues).

As examples of how these NGOs work synergistically through community-based initiatives and governance engagement, we highlight the Coalition's Community Supported Agriculture (CSA) Supper Club program and Ecology

North's food-related waste reduction efforts. The Supper Clubs use local food from the Yellowknife Farmers Market to teach cooking classes with young families and at-risk youth in Yellowknife. The Supper Clubs promote connections and knowledge sharing among their participants, local food producers, a local nutritionist, and other food-related community organizations (YKFM, 2017a). Similarly, Ecology North along with other actors, including the Yellowknife Farmers Market, have successfully encouraged businesses and events in Yellowknife to decrease the use of food packaging and to switch from single-use to compostable food packaging (Panza-Beltrandi, 2018; YKFM, 2017b). They achieved this change through public awareness campaigns, educating restaurants and other businesses about switching to compostable products and working together with the City of Yellowknife on their centralized composting program, and by promoting a Sustainable Event Checklist that encourages minimizing and using compostable food packaging (City of Yellowknife, 2018b). These types of on-the-ground initiatives enable these NGOs to work with various segments of the public to directly and indirectly shape and gain feedback on aspects of their visions for comprehensive and collaborative food systems governance in the NWT. Through such work, they also build public support that enhances their legitimacy when they engage in formal policy advocacy, as we saw in the case of the Coalition's effects on Yellowknife's municipal council in particular.

In sum, Ecology North and the Coalition demonstrate several internal characteristics that allow them to influence food governance processes, such as their ability to reframe municipal and territorial conversations about food by making the connections between knowledge, action, and policy. At the same time, external circumstances in the NWT also affect their ability to have an impact.

Political opportunity structure: External opportunities and constraints

How does the political opportunity structure in Yellowknife and NWT lend itself to NGO involvement in food governance, and what constraints might it present? We argue that one of the key dynamics of the political opportunity structure of the NWT is that it is both "heavily" and "lightly" governed. By heavily governed, we mean that there are many jurisdictional layers in the NWT, from municipal/community, territorial and federal governments, to various Indigenous authorities, including First Nations leadership, comprehensive and specific land claim agreements (between the federal and territorial governments and Indigenous peoples), and Indigenous self-government. An aspect of this heavy governance is also the siloed nature of food system-related sectors at the territorial level. Agriculture, fishing, and food-related timber-forest products are the mandate of the Department of Industry, Tourism and Investment (GNWT ITI, n.d.), whereas hunted, trapped, and gathered foods are housed within the Department of Environment and Natural Resources (GNWT ENR, n.d.). Further, food-related health concerns lie within the Department of Health and Social Services, including the Anti-Poverty Advisory Committee, to which many food security concerns are referred (GNWT HSS, n.d). When looking to engage in food systems governance, NGOs must navigate the multiple jurisdictional interests in the NWT as well as the internal divisions within them. At the same time, the NWT is lightly governed in certain ways. A low population density as well as the recentness of land claim negotiations and devolution mean that there is a relative newness to governance bodies in the NWT as well as limited capacity and resources for governing this vast territory. Together, these dynamics lead to opportunities and challenges for civil society engagement in governance. Below we highlight specific examples from the work of Ecology North and the Coalition to show how they navigate this governance context.

The heavily governed aspects of the NWT result in the need for NGOs working on governance issues to engage across multiple contexts. With their flexibility, NGOs, in general, are able to do this well by building inclusive communities around defined policy objectives (Jasanoff, 1997). Both Ecology North and the Coalition demonstrate what this looks like in practice, though they are at different stages in their work. Further, both NGOs engage Indigenous authorities and community members in these policy communities and in doing so seek to support Indigenous self-determination (however, it is not always clear how effective this engagement is, as we discuss below). In Yellowknife, the Coalition is deliberately drawn from multiple food system actors, from food producers, harvesters, foragers, and businesses to health practitioners, members of the YKDFN community, and organizations in Yellowknife interested in food security issues. The work of building the Charter was about developing both a shared vision, and about developing collaboration among the partners in the Coalition (YKFM, 2016). Similarly, Ecology North is now championing the NWT Food Network,[6] a new territorial-wide organization that brings together food system actors and community leaders to provide a platform for their voices in territorial and federal food policy as well as supporting new integrated food systems' initiatives in the NWT. The NWT Food Network is in the early stages of building collaborations and has been able to relatively quickly bring together some local food producers and First Nation leaders (Ecology North, 2018). As well, discussions with GNWT officials have begun, yet the comprehensive, collaborative vision for the NWT Food Network faces challenges with the siloed nature of the territorial government departments. Other challenges Ecology North (including the NWT Food Network) and the Coalition face include navigating how the heavily and lightly governed aspects of the political opportunity structure play out in varied ways across the different forms of food provisioning.

Hunting is relatively heavily governed with many structures surrounding it. The traditional governance of NWT's Indigenous peoples is an active part of the mesh of governing structures. These traditional governance systems include practices and institutions for managing the land as a common-pool resource (Parlee & McMillan, 2013; Parlee & Berkes, 2006; Parlee et al, 2005). On the territorial level, the Department of Environment and Natural Resource is responsible for the implementation and enforcement of the Wildlife Act, as well as providing resources to community-based Local Wildlife Committees for supporting local harvesting activities. Collaborative management of wildlife[7] is shared with Indigenous Renewable Resources Boards, as part of the settlement of land claim agreements. In non-settlement regions, management includes input and involvement by Indigenous organizations in wildlife management (GNWT ENR, n.d.). This highlights the multiple structures surrounding the governance of hunting in the NWT as well as the openness of the structures to Indigenous collaborative governance. However, it can be complicated for NGOs to know how to engage in this complex context. For example, the Food Network struggles with the questions of: If there are already so many structures surrounding hunting, is the Network needed (or wanted) to support Indigenous hunters? Could liaisons be an option? If so, who should the Network liaise with (Ecology North, 2018)?

The lightly governed aspects of the political opportunity structure in the NWT mean that governments, at times, welcome new capacity offered by civil society. However, this creates challenges when capacity and resources are minimal on both the side of governments and NGOs. For example, agriculture in the NWT currently has a limited governance structure. At the territorial level, the *Northwest Territories Agriculture Strategy: The Business of Food: A Food Production Plan 2017–2022* is a recent development created in the absence of any other agriculture policy. This strategy seeks to increase economic opportunities in the agriculture sector (GNWT ITI, 2017). While the Agriculture Strategy is an important policy development, the limited scope of agricultural governance to only commercial ventures means that civil society has a narrow opening to engage with government in agriculture. As a result, when meeting with GNWT officials regarding the NWT Food Network, there has been encouragement on the part of government for an agriculture sector interest group, but not for other parts of the comprehensive food systems vision.

At the municipal scale, food systems policy is lightly governed as it is not part of the mandate for that jurisdiction (Community Planning and Development Act, 2011), therefore any initiatives are at the discretion of the community/municipality. The first evidence that the Yellowknife context is open to new actors is found in the fact that the Coalition has a city councillor attend their meetings and the Coalition has worked with this councillor to create a link to City staff to discuss details of the local food strategy. This connection created an ally among the councillors, which provided clout and information about how to frame the Coalition's request for a Local Food Strategy in the second presentation made to council in 2017 (City of Yellowknife, 2017). There are constraints at the Yellowknife level, however. In their presentations to council, the Coalition outlined their vision of a comprehensive food strategy, touching on agriculture, but also Indigenous practices of hunting, fishing, gathering, all framed from a food systems perspective (City of Yellowknife, 2016a, 2017). To date, it appears that the City is not seeing a food strategy in this way. Discussions by the Coalition with City staff and the *Council's Goals and Objectives 2016–2018* (City of Yellowknife, 2016b) have highlighted that it will only be producing an *agricultural* strategy. Further, the Coalition would ideally have a comprehensive strategy created using a collaborative governance model, with financial support from the City for the Coalition to facilitate representation from across the food system. This funding has not been forthcoming (yet), and as a volunteer-based organization, the Coalition does not have sufficient capacity to take on a large role in the creation of a strategy.

Food system co-governance?

Given the evidence of their impact on food-related governance on multiple levels, have these NGOs been able to exert structural power to create co-governance mechanisms that include their active participation? We frame our response to this question in terms of whether the key co-governance dynamics proposed by Emerson et al (2011) are present in the work of these two NGOs: drivers, quality principled engagement, shared motivation, and sustained capacity for joint action.

First, of the four drivers of co-governance proposed by Emerson et al (2011) (government leadership, consequential incentives, interdependence, and uncertainty), only one of these is seen clearly in both contexts. At both the territorial and municipal level, we find a context of uncertainty as a result of a changing climate and the need to address the deep challenges of food insecurity in ways that are culturally appropriate and economically sustainable. Another common driver for co-governance, the necessity of interdependence among the relevant actors, is also arguably present, and is recognized at the municipal level. However, at neither municipal nor the territorial level do we find government leadership that is actively seeking to build co-governance with civil society (yet), nor do we see any incentives to formally include civil society organizations in co-governance.

Second, despite the presence of only one full and some partial drivers for co-governance, some quality principled engagement among key actors has been generated, through the partnerships and interactions with the state mentioned above. This has helped to build professional relationships as well as examples of trust and cooperation among the NGOs and state actors, which are important factors to building and sustaining co-governance arrangements (Addy et al, 2014). Still, that engagement remains nascent and has not always been explicitly about creating collaborative governance. It is too early to know if the engagement we see will produce more formal collaborative governance efforts.

Third, has a shared motivation, fostered through mutual understanding and shared commitment, been generated? If anything, it appears that both levels of governments are responding well to these NGOs on the topic of agricultural policy, but perhaps less so on a fully integrated policy response to food. At the municipal level, the focus on just an agriculture strategy from the City, instead of a more comprehensive food systems approach, shows that there may not be mutual understanding between the two groups or a lack of shared commitment. At the territorial level, there is a shared commitment between civil society and the state for bolstering agriculture as an economic driver in the NWT, however there does not seem to be a mutual understanding of how to achieve this. As an example, using a comprehensive food systems governance approach that works with agriculture, hunting, fishing, gathering, and imported foods to address economic development (among other objectives) is still a foreign concept to a siloed government.

The fourth dynamic of co-governance is the creation of sustained capacity for joint action through a "virtuous cycle" of principled engagement and shared motivation that stimulates the development of institutional arrangements, leadership, knowledge, and resources. These processes are still so new that we just do not know if this virtuous cycle will emerge. If it does, we speculate that it may come in the area of agricultural policy, and then grow outwards from there, but the barriers mentioned above will need to be addressed. Importantly, the interactions of Ecology North and the Coalition with the state are new. More instances of cooperation are clearly needed to strengthen these relationships (Addy et al, 2014), including more opportunities to pull all parties outside their comfort zones in productive ways.

Our overall assessment is that we do not see co-governance yet. Both governments still have the structural power in governance practices. They have been open to these NGOs but have not formally opened up governance processes, despite calls from these civil society actors to do so. In terms of the governance continuum presented in Chapter 1, we are clearly still at multi-stakeholderism. There is NGO interest in co-governance and there are the drivers of uncertainty and some principled engagement with the state, which shows promise of potential. However, we certainly do not see strong evidence of co-governance as regular practice yet.

Conclusion

This chapter examined the emerging role of the Yellowknife Food Charter Coalition and Ecology North in food governance processes in the NWT.

We considered how these NGOs are working to shape municipal and territorial governance, and whether that influence is resulting in the collaborative governance (or co-governance) role which these NGOs seek.

Building on social movement theory, we found that these NGOs are effective at combining new frames, coalition building, and working synergistically on policy and community initiatives. However, the political opportunity structure has some constraints. It seems particularly open to a municipal agricultural policy, but no evidence yet that the municipality will think of a food strategy in more comprehensive, systemic, terms. At the territorial level, Ecology North is moving into the coalition-building phase with its territorial network, hoping to have a growing impact. The political opportunity structure is open from the point of view of there being lots of space within the NWT to influence certain policies (again, especially agriculture), but that structure may also impose constraints as other areas relevant to food, hunting, and resource management are already heavily governed by a range of invested actors (though some of that is through Indigenous co-management regimes).

We drew on Emerson et al's (2011) work on the dynamics of co-governance, which proposed that co-governance in practice involves certain types of key drivers, quality principled engagement, shared motivation, and a shared capacity for joint action. Looking for these qualities in NGO engagement with food governance in the NWT leads us to identify some opportunities for co-governance to emerge, but greater engagement and trust-building among key actors must occur first.

Acknowledgements

We would like to acknowledge the Social Sciences and Humanities Council of Canada (SSHRC) for supporting this research through the FLEdGE Partnership Grant. The primary author extends her sincere thanks to the Yellowknife Food Charter Coalition and Ecology North. Their dedication and drive for just and sustainable food systems in Yellowknife and throughout the NWT is a constant source of inspiration.

Notes

1 Compared to food systems that are largely based in agriculture, food systems that are rooted in hunting, fishing, gathering, and imported foods, such as those in the NWT, are under-studied (and arguably under-valued). This chapter helps to fill that gap and adds diversity to foods systems literature.
2 FLEdGE is a research and knowledge sharing partnership, committed to fostering food systems that are socially just, ecologically regenerative, economically localized and that engage citizens. For more information, please see www.fledgeresearch.ca
3 From the 1880s to the 1990s, residential schools were part of an extensive school system that was set up by the Canadian government and administered by churches with the explicit objective of assimilating Indigenous children into Euro-Canadian

society and Christian ways of living. The system forcibly separated children from their families for extended periods of time and forbade them to acknowledge their Indigenous heritage and culture or to speak their own languages. Former students have spoken of horrendous physical, sexual, emotional, and psychological abuse at residential schools.

4 For more information of the Yellowknives Dene First Nation, please refer to www.ykdene.com

5 For information on the North Slave Metis Alliance, please refer to www.nsma.net

6 For information on the NWT Food Network, please refer to www.NWTfoodnetwork.ca. The primary author is closely involved in the creation of this organization.

7 Collaborative management of wildlife in the NWT is not without its challenges and power imbalances. Parlee et al (2018) argue that there is a "policy–science gap" in the NWT and the Yukon when it comes to wildlife management that excuses the detrimental environmental disturbances of resource extraction on wildlife and places bans on harvesters as the primary form of management. This gap has parallels to historic governance systems that were "explicitly about advancing private interests in northern lands and resources at the expense of Indigenous cultures and livelihoods" (p. 1).

References

Addy, N. A., Poirier, A., Blouin, C., Drager, N., & Dube, L. (2014) Whole-of-society approach for public health policymaking: a case study of polycentric governance from Quebec, Canada. *Annals of the New York Academy of Sciences*, December, 216–229.

Chevalier, B. & Buckles, D. (2013) *Participatory action research: theory and methods for engaged inquiry*. New York, Routledge.

City of Yellowknife (2015) *Municipal services committee report*. June 23. City of Yellowknife.

City of Yellowknife (2016a) *Municipal services committee report*. August 22. City of Yellowknife.

City of Yellowknife (2016b) *Council's goals and objectives 2016–2018*. City of Yellowknife.

City of Yellowknife (2017) *Municipal services committee report*. August 14. City of Yellowknife.

City of Yellowknife (2018a) *Citizen budget results*. OpenNorth. City of Yellowknife.

City of Yellowknife (2018b) *Yellowknife sustainable even checklist*. Available at: www.yellowknife.ca/en/Modules/News/

Council of Canadian Academies (2014) *Aboriginal food security in northern Canada: an assessment of the state of knowledge*. The Expert Panel on the State of Knowledge of Food Security in Northern Canada, Council of Canadian Academies. Ottawa.

Dene Nation, The (2016) *History*. Available at: www.denenation.ca/history/

Ecology North (2017a) *About us*. Available at: http://ecologynorth.ca/about-us/

Ecology North (2017b) *Food systems visioning*. Ecology North.

Ecology North (2018) *A sustainable food organization for the NWT? Meeting report*. Ecology North.

Emerson, K., Nabatchi, T., & Balogh, S. (2011) An integrative framework for collaborative governance. *Journal of Public Administration Research*, 22(1), 1–29.

Foucault, M. (1978) *The history of sexuality: an introduction*. New York, Random House.

Friedmann, H. (1992) Distance and durability: shaky foundations of the world food economy. *Third World Quarterly*, 13(2), 371–383.

GNWT ENR (Government of Northwest Territories, Department of Environment and Natural Resources) (n.d.) *Environment and natural resources programs and services*. Available at: www.enr.gov.nt.ca/en/services

GNWT HSS (Government of Northwest Territories, Department of Health and Social Services) (n.d.) *Current anti-poverty advisory committee*. Available at: www.antipovertynwt.ca/en/current-anti-poverty-advisory-committee

GNWT HSS (2011) *Northwest Territories health status report*. Available at: www.hss.gov.nt.ca/sites/hss/files/nwt-health-status-report.pdf

GNWT HSS (2017) *Arsenic in lake water around Yellowknife*. August 1. Available at: www.hss.gov.nt.ca/en/newsroom/arsenic-lake-water-around-yellowknife

GNWT ITI (Government of Northwest Territories, Department of Industry Tourism and Investment) (n.d.) *Industry, tourism, investment programs and services*. Available at: www.iti.gov.nt.ca/en/services

GNWT ITI (2017) *Northwest Territories agriculture strategy: the business of food: a food production plan 2017–2022*. Available at: www.iti.gov.nt.ca/sites/iti/files/agriculture_strategy.pdf

Jasanoff, S. (1997) NGOs and the environment: from knowledge to action. *Third World Quarterly*, 18(3), 579–594.

Kindon, S., Pain, R., & Kesby, M. (2007) *Participatory action research approaches and methods: connecting people, participation and place*. New York, Routledge.

Kooiman, J. (2003) *Governing as governance*. London, Sage.

Leahey, A. (1954) Soil and agricultural problems in subarctic and arctic Canada. *Arctic*, 7(3/4), 249–254.

Lutra Associates Ltd (2010) *Yellowknife, Ndilo and Dettah food system assessment and community food action plan*. Northwest Territories and Nunavut Public Health Association.

McAdam, D., McCarthy, J. D., & Zald, M. N. (eds) (1996) *Comparative perspectives on social movements: political opportunities, mobilizing structures and cultural framings*. Cambridge, Cambridge University Press.

NWT Bureau of Statistics (2014a) *Aboriginal persons who hunted & fished during 2013, by community*. NWT Bureau of Statistics.

NWT Bureau of Statistics (2014b) *Persons who gathered berries during 2013, by community*. NWT Bureau of Statistics.

O'Reilly, K. (2015) Liability, legacy, and perpetual care: government ownership and management of the Giant Mine, 1999–2015. In: Sandlos, J. & Keeling, A. (eds) *Mining and communities in Northern Canada: history, politics, and memory*. Calgary, University of Calgary Press, pp. 341–367.

Paquet, G. & Wilson, C. (2011) Collaborative co-governance as inquiring systems. *Optimum Online*, 41(2), 2–16.

Panza-Beltrandi, G. (2018) Some Yellowknife bars choosing to ditch plastic straws. *CBC News*. April 23. Available at: www.cbc.ca/news/canada/north/yellowknife-straw-ban-bars-1.4629800

Parlee, B. & Berkes, F. (2006) Indigenous knowledge of ecological variability and commons management: a case study on berry harvesting from northern Canada. *Human Ecology*, 34(4), 515–528.

Parlee, B. & McMillan, R. (2013) Dene hunting organization in Fort Good Hope, Northwest Territories: "ways we help each other and share what we can". *Arctic*, 66(4), 435–447.

Parlee, B., Berkes, F., & Council, T. G. (2005) Health of the land, health of the people: a case study on Gwich'in berry harvesting in northern Canada. *EcoHealth*, 2(2), 127–137.

Parlee, B., Sandlos, J., & Natcher, D. (2018) Undermining subsistence: barren-ground caribou in a "tragedy of open access". *Science Advances*, 4(2), 1–14.

Receveur, O., Boulay, M., & Kuhnlein, H. V. (1997). Decreasing traditional food use affects diet quality for adult Dene/Metis in 16 communities of the Canadian Northwest Territories. *The Journal of Nutrition*, 127(11), 2179–2186.

Sandlos, J. (2007) *Hunters at the margin: native people and wildlife conservation in the Northwest Territories*. Vancouver, University of British Colombia Press.

Sandlos, J. (2015) "A mix of the good and the bad": community memory and the Pine Point Mine. In: Sandlos, J. & Keeling, A. (eds) *Mining and communities in Northern Canada: history, politics, and memory*. Calgary, University of Calgary Press, pp. 341–367.

Sandlos, J. & Keeling, A. (2016) Toxic legacies, slow violence, and environmental injustice at Giant Mine, Northwest Territories. *The Northern Review*, 42, 7–21.

Sharma, S., De Roose, E., Cao, X., Pokiak, A., Gittelsohn, J., & Corriveau, A. (2009) Dietary intake in a population undergoing a rapid transition in diet and lifestyle. *Canadian Journal of Public Health*, 100(6), 442–448.

Sheehy, T., Kohlahdooz, F., Schaefer, S. E., Douglas, D. N., Corriveau, A., & Sharma, S. (2014) Traditional food patterns are associated with better diet quality and improved dietary adequacy in Aboriginal peoples in the Northwest Territories, Canada. *Journal of Human Nutrition and Dietetics*, 28(3), 262–271.

Schurman, R. (2004) Fighting "frankenfoods": industry opportunity structures and the efficacy of the anti-biotech movement in western Europe. *Social Problems*, 51(2), 243–268.

Schurman, R. & Munro, W. (2009) Targeting capital: a cultural economy approach to understanding the efficacy of two anti-genetic engineering movements. *AJS*, 115(1), 155–202.

Spring, A. (2018) Capitals, climate change and food security: building sustainable food systems in northern Canadian Indigenous communities (Unpublished doctoral dissertation). Wilfrid Laurier University, Canada.

Tarasuk, V., Mitchell, A., & Dachner, N. (2016) *Household food insecurity in Canada, 2014*. Toronto, PROOF.

Tarrow, S. (1998) *Power in movement: social movements and contentious politics*. Cambridge, Cambridge University Press

Voss, J. P., Bauknecht, D., & Kemp, R. (eds) (2006) *Reflexive governance for sustainable development*. London, Edward Elgar.

Wapner, P. (1995) Politics beyond the state: environmental activism and world civic politics. *World Politics*, 47(3), 311–340.

Wesche, S. & Chan, H. M. (2010) Adapting to the impacts of climate change on food security among Inuit in the western Canadian Arctic. *EcoHealth*, 7(3), 361–373.

Wesche, S., O'Hare-Gordon, M. A., Robidoux, M. A., & Mason, C. W. (2016) Land-based programs in the Northwest Territories: building Indigenous food security and well-being from the ground up. *Canadian Food Studies*, 3(2), 23–48.

YKCGC (Yellowknife Community Garden Collective) (2017) *About us*. Available at: www.ykgardencollective.org/food-security

YKFM (Yellowknife Farmers' Market) (2016) *The Yellowknife food charter*. Available at: www.yellowknifefarmersmarket.ca/foodcharter/

YKFM (2017a) *YK CSA supper club 2017 final report*. Yellowknife Farmers' Market.

YKFM (2017b) *Waste reduction and composting program*. Available at: www.yellowknifefarmersmarket.ca/waste-reduction-and-composting-program/

Zotor, F., Sheehy, T., Lupu, M., Kolahdooz, F., Corriveau, A., & Sharma, S. (2012) Frequency of consumption of foods and beverages by Inuvialuit adults in Northwest Territories, arctic Canada. *International Journal of Food Sciences and Nutrition*, 63(7), 782–789.

3 Searching for fit?

Institution building and local action for food system change in Dunedin, New Zealand

Philippa MacKay and Sean Connelly

Introduction

The relationship between food and concerns about health, the environment, local economies, community development, and social justice provide multiple opportunities to access a range of resources to support food system change (Morgan, 2015; Campbell et al, 2013; Levkoe, 2011; Sherriff, 2008). Yet, these diverse opportunities also highlight the important role of food governance processes to resolve tensions, conflicts, and contradictions that often arise from competing approaches, norms, and values about the meaning of food system change (Jarosz, 2014; Guthman, 2008). For Goodman et al (2012), governance processes that are reflexive and that embrace contradictions and difference are critical to creating the democratic space for food system transformation. Similarly, Hassanein (2003) stresses the need to actively create opportunities for equal and effective opportunities to participate in decisions about the food system.

As discussed in the introductory chapter of this book, reflecting on the implications of different forms of governance arrangements, the values that underpin them (and how they are negotiated), and the risks and benefits associated with active participation in governance relationships offers important insights for how food movements might harness governance processes to advance more sustainable food systems. This chapter presents specific insights from the attempts of a small group of alternative food system actors to broaden discussion about food issues by engaging directly with their local government. In a similar vein to the discussion of the Yellowknife Food Charter Coalition (Chapter 2) and Correns France (Chapter 9), these efforts can increase access to resources and generate more capacity to change the food system. However, they also highlight the need to create deliberate opportunities to reflect on the trade-offs, compromises, and conflicts that emerge in doing so.

The transformative potential of food, where "food is/has been a lever for change" (Blay-Palmer et al, 2014, p. 186) depends on robust and adaptive governance arrangements where the process of engaging in food politics involves creating deliberate places for diverse stakeholders to reflect on their practice and to engage in local food democracy. Creating opportunities for reflexive practice requires thinking of the politics of alternative food system-making as a process where diverse actors collectively work out the tensions and contradictions in place that can lead to a restructuring of not just food, but also broader social relations and values (Marsden, 2013; McClintock, 2013; DuPuis & Goodman, 2005). Activating food democracy transposes food initiatives beyond the "fetishization of food" (Guthman, 2011) by making the social relations of food production, distribution, and consumption transparent and open to political debates. These kinds of governance arrangements facilitate new models of citizenship in which there are opportunities for individual participation and collective community organizing.

In this chapter we explore these broad issues through the relationships, politics, and cultures of two co-evolving food system networks in Dunedin, New Zealand (NZ). The first network, Our Food Network Dunedin (OFN), is a self-described grassroots organization dedicated to stimulating the production, distribution, and consumption of local food, and in that way contribute to building a resilient and prosperous community. OFN recognized the importance of local council involvement to support food system change and were successful (along with other actors) in lobbying the Dunedin City Council to create a part-time position in 2015 to ensure that food issues were considered across council policies and in the broader community. Out of these new council resources emerged Good Food Dunedin (GFD), a council-led food initiative that exists to fill a food system governance void by bringing together diverse stakeholders who share a vision of transforming Dunedin into a thriving and sustainable food city. Up until recently, efforts dedicated to food system change in Dunedin were limited to discrete individual projects and there was no mechanism to identify, discuss, or reflect on the potential for more systematic efforts. We discuss the evolving process of local food system governance based on the efforts of these two co-existing networks to reflect on how diverse stakeholders are engaged in shaping the meaning and process of food system change.

Through an analysis of OFN and GFD, we argue that what was once a relatively small civil society effort to broaden discussion amongst existing alternative food system actors has resulted in strengthened engagement with the council to place food on the public agenda. As a result, problems and solutions related to food system change had to be reframed through negotiation and compromise. While the pragmatic trade-offs might appear to be less transformative, they have resulted in dramatically extending the reach and prominence of food system change as an issue for discussion in Dunedin.

This chapter is based on 24 interviews with a range of food system actors that were conducted in 2016 by one of the authors. The interviews were broadly focused on the challenges and limitations of reimagining the Dunedin food system. In addition, we have relied on the notes, meeting minutes, and reflections of the other author who has been involved, since 2013, in both of the networks discussed. The chapter begins by providing some context on food system governance in New Zealand.

Food system governance in New Zealand

Over the last three decades, the role and process of government and governance has changed dramatically in New Zealand, making it "a particularly interesting case because of its well-deserved reputation as a social democratic welfare state that went neoliberal with a vengeance in the mid-1980s" (Peet, 2012, p. 151). In the 1970s, the New Zealand economy and society was organized around an overtly welfare state, financed in large part by privileged access to the United Kingdom (UK) market for its agricultural products, and was characterized by high levels of state management and intervention. Global economic crises, increasing globalization, freer trade, the loss of preferred access to the UK market, and sectoral shifts in the economy in the 1980s resulted in the New Zealand government's decision to rationalize the welfare state and embrace neoliberalism (Connelly & Nel, 2016). These changes had profound effects on people and communities, as the closure of state-led primary industries and services had severe implications on local employment. In addition, NZ has the distinctive political-economic history of having transformed from one of the most protected and controlled economies in the OECD (Organization for Economic Co-operation and Development) and being a champion of social democracy to one of the least interventionist countries in the world (Conradson & Pawson, 2009; Challies & Murray, 2008). These changes have had profound and lasting implications on the perception and role of government in relation to society and the economy.

The new relationship of government to society and the economy is reflected in the national approach to food and agriculture policy. The transition to a neoliberal agricultural economy resulted in new private-sector driven governance mechanisms that placed privileged focus on intensification and maximizing export potential of key products such as dairy, sheep and beef, and fruit. Up until recently, food policy discussions were limited to issues of market access through trade agreements and intensification on the one hand and private-sector driven standards designed to satisfy the food quality demands of international high value retailers around issues such as sustainability and food safety on the other (Campbell, 2005). Issues of food security and food justice, linkages to local economic development, social engagement around food, concern for agricultural labour and environmental sustainability of food for local consumption have been pushed to the margins.

More recently, however, the absence of action at the national level has stimulated response at the local level. There is increasing local-level engagement with food policy through food policy councils, food charters, and local government involvement in food which brings formal institutional resources and governing capacity to support ongoing activities such as community gardens, buy-local campaigns, organic production, and alternative markets that have been pursued almost exclusively by civil society actors (Dwiartama & Piatti, 2016; Bartos, 2014). Yet, as Amin (2011) reminds us in the context of food sovereignty movements, these localized efforts are unlikely to transform food systems on their own; they must also collaborate, share knowledge and experience, and be engaged in the politics of food system change at multiple scales. As illustrated by Levkoe and Wilson (Chapter 5), efforts to connect at a national level can not only strengthen the voice and power of food movements beyond place, but they also provide the potential to use the reframing of national-level food policy as a means to galvanize further support in place through the sharing of resources, strategies, and mobilizing further interest.

The larger food policy context of NZ that has resulted from the roll-back and roll-out of neoliberalism (as discussed in the introductory chapter) has created room for civil society actors to play a greater role. However, many of these activities are locally focused, either on filling gaps that arise from the roll-back of the state (e.g. addressing access to food) or through the emergence of civil society engagement in new opportunities for contentious politics that directly challenge the direction the state has taken with regards to food system issues. While this space for food politics is emerging locally, little effort has been made by these civil society actors to date to make connections beyond places or to engage in national food policy discussions in NZ. OFN was part of a group of food networks from Auckland, Christchurch, and Wellington that considered the potential of organizing nationally in 2015 and 2016 with a goal of discussing how their collective work in place could contribute to assisting and connecting local food initiatives across NZ. This was seen as the potential starting point for a national local food network. However, due to resource constraints and more pressing priorities, efforts did not advance beyond those initial discussions.

The broader neoliberal context also shapes how decisions about food are made in Dunedin. Up until recently, the food governance vacuum meant that the starting point for alternative food systems was not a reform/transform debate, but rather making the case that there is scope and justification to do anything. For example, during a recent debate about the Good Food Dunedin Food Charter, a Dunedin councillor questioned the "waste of ratepayers money" when "supermarkets are doing a perfectly adequate job, are they not?" (Morris, 2018), highlighting the view that there is no need for local food policy or governance, as there is no policy problem that needs to be addressed. We highlight below how OFN and GFD emerged as two key actors making the case for deliberate local food system governance and how they used a combination of instrumental, structural, and discursive power (Clapp & Fuchs, 2009) to justify and create an emergent co-governance process that allows for multiple actors to negotiate food system change.

Our Food Network Dunedin

Dunedin is a small city with a population of 120,249 (Stats NZ, 2013), surrounded by a large agricultural hinterland in the South Island of NZ (see Figure 3.1). It is dominated by university with nearly 20,000 students.

Figure 3.1 Map of New Zealand
Source: Chris Garden (used with permission)

Dunedin has a long standing farmers' market, backyard producers, niche producers, FoodShare (a food rescue organization described in subsequent section), and organic and local food restaurants and retailers that occupy the food space in conjunction with export-oriented producers of dairy, sheep, beef, wine, and fruit, and foreign-owned supermarket chains. Dunedin is also characterized by close-knit neighbourhoods with community gardens and community orchards that have arisen out of concerns for health and nutrition, improved food access, and community resilience.

OFN has its roots in a series of local food forums organized by a small group of individuals that were held in 2012 and 2013 and were designed to bring together food producers, activists and advocates to explore the potential for stronger connections across the food system (see Figure 3.2 for timeline). OFN emerged from these food forums as an organization supported by over 400 members with an ad-hoc steering group that would take ideas forward, facilitate networking, and mobilize community support "to stimulate the production, distribution and consumption of local food and in that way contribute to the building of a resilient and prosperous community" (OFN, n.d.). Much of the focus of the initial food forums was on what the Dunedin City Council (DCC) could do to support local food.

OFN organized a subsequent food forum in 2014 in an effort to connect existing diverse initiatives, explore how resources might be shared, and to identify how individual initiatives might be supported to achieve a food system that is more just and sustainable. In contrast to the 2013 forum, the focus of discussion at the 2014 forum had shifted to identifying what actions members of the network could take, which is indicative of a shift in agency and a maturation of the network (Haylock & Connelly, 2018). At a "think tank" meeting subsequent to the 2014 food forum, member's prioritized four key projects that they could advance based on their level of commitment, knowledge, and capacity. They included 1) seed saving; 2) land share (both rural and urban); 3) food identification and labelling; and 4) communication and education projects. Since that time, OFN has been actively involved in local food advocacy, mobilizing submissions in response to DCC policy and planning processes, project support, and fundraising. OFN has been successful in obtaining grant funding to support a part-time "ideas-to-action" facilitator to advance key projects, funded

Figure 3.2 Timeline of activities

Source: Authors/Chris Garden (used with permission)

through a combination of DCC and community trust grants. In addition, an opportunity arose to expand an existing community fruit tree harvest project to increase food production, improve access to nutritious local food, and raise awareness about a range of food issues (such as healthy eating, food budgets, community development, environmental impacts, etc.) in neighbourhoods across the city. Working with organizations such as Well South (the main regional health promotion organization), Presbyterian Support Otago (the main foodbank provider in Otago), Heart Foundation, Enviroschools, and FoodShare (a food rescue organization described below), OFN's neighbourhood food harvest project supports communities to become more engaged in their own food systems (OFN, n.d.).

Good Food Dunedin

At the same time, DCC began identifying the ways that the council could be more deliberate with regards to supporting local food resilience, shaped in part by the impact of the Christchurch earthquakes in 2010/2011. In 2013, the Community Resilience Forum (a council working party established in 2010) decided to focus on food, and in late 2013 they provided a report to DCC that suggested how Dunedin could address a range of food issues and also identified opportunities for the council to incorporate local food considerations into planning and development processes (DCC, 2013). One of the results of that work (in part shaped by the outcomes of the 2013 food forum) was a clear recommendation that the DCC could play a more active role. In 2015, the DCC completed a policy analysis that explored different options of resourcing and supporting a more coordinated response to food issues in Dunedin and identified options to better engage with stakeholders to advance food resilience across the city (DCC, 2015). The initial policy analysis research was focused on economic opportunities and climate change adaptation as a means to justify council involvement, although the focus eventually broadened out to include issues of food security and access to healthy food. In 2016, DCC committed resources for a part-time food resilience business advisor in Enterprise Dunedin, an organization within council responsible for economic development, promotion, and marketing for the city, to facilitate the GFD project.

According to DCC, food resilience consists of supporting thriving local food producers and businesses, protecting the food production environment, supporting better environmental outcomes, enabling communities to grow their own food, and ensuring an equitable supply of healthy and affordable food for residents (DCC, n.d.). OFN was supportive of council's role in addressing food resilience, but also had reservations about the business development focus and worried that the food agenda could be captured by more powerful actors (Haylock & Connelly, 2018). One of the first tasks of GFD was to set up an advisory group, the Good Food Alliance (GFA), whose role was to provide an opportunity for a broad range of food stakeholders in the city, including OFN, industry groups, charitable organizations, public health officials, growers, farmers' market members, retailers, restaurants and café owners, and researchers (GFD, n.d.a).

Over the following year, GFD (with the support of the GFA) led the development of the Dunedin Food Charter as a way of enabling partnerships of public, private, and community organizations to work together, using food as a focal point for change. The Charter commits signatories to working collaboratively to ensure that the vision of a thriving, food-resilient city that is good for people, the city, and the planet becomes a reality (GFD, n.d.b). The Dunedin Good Food Charter was approved by the council in 2018 and provides formal institutional and governance mandate for council leadership in addressing food system change in Dunedin.

Having described the broader context and the formation of OFN and GFD, we now turn to individual projects and organizations in the charity food sector that serve as examples of the diversity of ways that the problem of food access is framed. As discussed in Chapter 1, governance processes depend on a common organizing frame and ideological orientation to food system change. The purpose of these examples from the charity food sector is to illustrate how framings of food access shape participation in governance processes, which in turn shape the range of actors, the resources they bring, and their ability to collectively contribute to food system change in a more systematic way. This challenge of framing is not unique to the charity food sector, but is illustrative of the broader issues that result from the absence of formal governance process at the local level and the limited degree to which community and council efforts, such as OFN and GFD, have been able to articulate framings of food system problems and solutions in a meaningful way for these organizations.

Potential for food system governance?

Food networks such as OFN and GFD reflect a wide range of food actor perspectives. Individual groups are motivated to address food system problems for a range of reasons, including supporting local production, local economies, food access and food security, environmental, health, and wellbeing concerns. These diverse concerns reflect a combination of individual and group priorities, capacities, values, and approaches that together frame specific responses to food problems. In this section, we provide an example of the types of issues that result from complimentary and divergent values that underpin different framings that might be loosely categorized as being concerned about food access and security. We illustrate how, in an area where there is agreement that hunger is the common problem, diverse (and sometimes contradictory) responses emerge, highlighting the need for innovative governance mechanisms that provide opportunities to work out tensions around competing values for food system change. Despite the emergence of OFN and GFD, the three examples discussed below illustrate how the absence of deliberate discussion of the way problems are framed provides little clarity or discussion of how these individual efforts might be better supported to achieve their particular common goals of improving food security, or how their efforts might contribute to broader efforts of food system change in Dunedin.

Presbyterian Support Otago runs one of the foodbanks in Dunedin and it was set up to combine the need to provide both food and social service assistance to those in need. As one research participant explains:

> there is a lot of things that go on behind the scenes of people needing food and those are the things around poverty and issues around other things that concern us and interest us and having the foodbank and giving people food is one thing, but it's about what is enabling them or what is keeping them in that position that they actually need to come to a food bank in the first place

The approach by the foodbank organization is to tackle hunger in conjunction with other serious social issues that may be occurring within a family or in the community. Access to food parcels is linked to referrals from social service agencies that match food handouts with services such as household budgeting support, access to welfare benefits, or addiction and mental health support. Food in this instance is seen as a symptom of larger societal problems that need to be addressed and as a result, the foodbank works with a network of actors dominated by social services and charities that can assist and support foodbank users to improve their food security.

Another example of an organization working to improve food access is the All Saints' Fruit and Vege weekly food box program. The program is spearheaded by volunteers at a local church that buy blemished or otherwise leftover fruits and vegetables in bulk directly from commercial wholesalers in Christchurch (located ~360 km further north) and through volunteers, repackage the food into weekly food boxes for distribution to individuals or families. The key values underpinning this initiative not only aim to provide access to fresh and healthy food but have also been shaped by the concept of creating empowerment and dignity. As one research participant states

> [at] All Saints, this is not local food – fruit and veggies – this is coming from the market, but the difference is they charge three, six, and twelve dollars [per box]. It's very minimal, but their philosophy . . . is, if you don't put a cost, and people can't contribute to buying their food, they are getting it, then you're disempowered . . . so the key word that [is used] is dignity.

For All Saints, the primary motivation for their activities is reducing the cost of food to improve access to fresh fruit and vegetables. They rely on a large network of volunteers that are committed to contributing their sweat equity to purchase, transport and repackage food in the interest of reducing costs.

Finally, FoodShare has a similar aim to reduce food insecurity, but addresses the problem from the perspective of addressing the environmental problem of food waste. As one research participant stated "[the founder] saw two big problems, one of them being food insecurity within our city . . . and then she also saw that perfectly good food was being wasted". FoodShare is a food rescue venture that takes expired or excess food items from supermarkets, restaurants, and other food businesses and creates an assortment of free food packages that are delivered around the Dunedin community. FoodShare has successfully addressed a waste problem for local businesses and as a result has developed strong networks that include large supermarket chains on the one hand, to farmers' market vendors on the other, while addressing City Council concerns about diverting waste from landfills. In addition, food that would otherwise be thrown away feeds hungry people. The values associated with reducing food waste can also be understood from a variety of viewpoints. When asked about the drivers behind the DCC's support for FoodShare, one key informant suggested that the launch of a national food waste campaign 'Love Food Hate Waste' played a key role. They stated, "for me it's just reducing it [food waste], for other people it's about saving money, for other people it could be the carbon emission aspect to it". This informant further explained that the FoodShare model is exceptional because it weighs all food that comes in and out of its premises. It then records the statistics which is part of its success, as they can see how much food has been diverted from landfill and how many meals they have given to those in need of food packages. Considering the success and popularity of the FoodShare model (which has since expanded to Auckland under the organization's rebranded name of KiwiHarvest), approaches that can be quantified are a valuable information resource and have been effective at raising the credibility of the organization and demonstrating to funders the value of continued support. Local supermarkets and other food businesses, volunteers, and DCC, who provide the premises for FoodShare to physically work in, all contribute to their day-to-day operations. FoodShare is able to demonstrate its success in terms of increasing the amount of food diverted from waste streams and feeding more hungry people year over year, and deliberately limit their activities only to bringing those two priorities together to focus their resources and maximize their impact.

Despite all agreeing that food security is a problem that needs to be addressed, these three very different approaches to the problem (i.e. social service provision, increasing the availability of cheap food, and reducing food waste) each use different discursive framings of hunger as a means to engage with broader institutional structures to unlock resources to pursue their goals. For example, Presbyterian Support Otago frames hunger primarily as an individual problem, and in doing so is able to access the support of other social service agencies and government to assist individuals to cope with food insecurity. As discussed in the section below, however, framing hunger and food access in this manner can limit new methods and partnerships with stakeholders. Specifically, it creates challenges for foodbanks and food policy actors who wish to work more effectively together to address food poverty (Booth & Whelan, 2014). Similarly, FoodShare's framing of food security as linkages between hunger and food waste unlocks resources from supermarkets, corporate donors, and local government. However, in the absence of governance mechanisms that provide a venue for reflection on their individual practices, each of these organizations remains focused on their piece of the food security problem, and representatives of these organization are often reluctant to be engaged with broader food system change due (i.e. through OFN or GFD) to their limited resources. While they are able to exercise discursive power to unlock resources from other actors, they are not able to shape a broader food security agenda related to living wages, increased social and mental health services, or the right to food on their own.

By creating a venue to explicitly discuss food issues more broadly, the opportunity exists to bring together not just these three organizations, but also their diverse networks and resources to achieve broader transformational goals that might include mobilizing around the need to improve social benefits, increase minimum wages, or to transform supermarket practices that persistently view food as disposable. While there are general commitments to food rights and social justice, in the case of OFN, and food that is good for people in the case of GFD, discussions of how to operationalize these commitments have yet to materialize. As a result, organizations such as FoodShare, Presbyterian Support, and All Saints' Fruit and Vege box have been unable to translate their resources and networks into broader discussions about the future of the Dunedin food system, and nor have they been able to engage in discussions that might address the tensions and contradictions that are embedded in their individual responses. It is hoped that GFD will provide the mechanism for these discussions to occur, although as we will discuss below, the competing framings of food "issues" in order to access resources for particular projects make it difficult for diverse organizations to see the value of working more closely together. For example, the framing of food issues around local food (for OFN) and community resilience (for GFD) limit the value these charity food organizations see in being more actively engaged in these forums. In the section that follows, we examine in more detail the emergence of GFD and discuss the potential to shape the transformation of Dunedin's food system.

Governance relationships: Instrumental, structural, and discursive power and resources

The emergence of food system governance mechanisms such as OFN and GFD represent nascent attempts to collectively frame issues, advocate for solutions, and contribute to decision-making that might transform food systems. In this section, we analyse how a diversity of voices have come together in Dunedin to change aspects of the food systems. We discuss how the individuals and groups, including OFN, have used discursive power to place food on the agenda and how resources have been converted through the emergence of GFD as a mechanism to increase access to the instrumental and structural power of DCC by creating a deliberate space for food system governance to occur. Yet, despite language that makes reference to issues of food justice, equity, and food security in their overarching purpose, the discursive framing of food issues as primarily a local production problem or an economic development opportunity has limited the inclusiveness of efforts for food system change. Through a process of negotiation and compromise, the discursive framings of food system change have shifted as institutional voices gain prominence.

The various voices in Dunedin that have come to together to address food system change have united around similar core values such as local food and food resilience. The aftermath of the 2010/2011 Christchurch earthquakes shaped initial discourse on food resilience (Haylock & Connelly, in press). The disaster saw damaged warehouses and stocks of commercial and retail food ruined, resulting in residents being unable to purchase or acquire basic food items such as bread and milk (Paci-Green & Berardi, 2015). The significance of that event has played a role both in emphasizing and elevating the important need for action around food resilience and has highlighted the fragility of the food system in Dunedin, where supermarket chains are reliant on food trucked in from warehouses in Christchurch, which is located some 360 kilometres away (DCC, 2013).

DCC used their institutional resources and the structural power of council sub-committees to put food on the agenda through the formation of the Community Resilience Forum. DCC provided time, resources, and money to allow an employee to gather information, suggest actionable steps, and provide a set of recommendations on how the council might engage with food. The policy paper *Community Resilience Forum Recommendations Report On Food* (DCC, 2013) was produced to assist with conversations and decision-making around the future of the food system and its governance in Dunedin.

In parallel, engagements between diverse actors had been occurring informally, as, for example, local food producers from the farmers' market interacted with processing entities, food distribution, and consumers on a weekly basis (Dwiartama & Piatti, 2016). The same is true for the network of actors involved with community gardens, foodbanks, or food waste organizations, where organizations worked together to ensure that individuals who were food insecure were made aware of the range of options through which they could improve their access to healthy and nutritious food through community gardens, foodbanks, FoodShare, and through the vegetable box scheme. Each of the organizations discussed above provided access to different networks and resources. For example, the foodbank was primarily networked with other social service agencies, offering support for access to housing, social benefits, and health services. The All Saints vegetable box program provided a mechanism for participants to access cheaper food, but also social connections to other community members through volunteering opportunities and empowerment. Finally, FoodShare's networks developed linkages between supermarkets, the farmers' market, cafes, and restaurants and the charity food sector by increasing the availability of food donations for those organizations. Collectively, they represented a growing, albeit uncoordinated, conversation that food system changes needed to occur, and that civil society groups were willing to invest their time and energy into developing governance mechanisms for those discussions to occur.

While OFN played an important role as the vehicle through which many civil society organizations invested their energy to create opportunities to discuss food system change, it was also actively involved in creating and supporting food initiatives in the city, which opened up a range of new knowledge, capacity, and resources from the council that had not previously been allocated to discussions about food. Rather than being seen as multiple individual food projects, OFN provided a collective and coordinated voice that provided council staff and politicians the opportunity to claim that there was broad community support and legitimacy that justified investment of council funds in GFD. The discursive power mobilized by OFN provided legitimacy for food as an issue to be discussed locally and also provided opportunities and motivation for people in civil society to work together around food and to contribute their dispersed instrumental power more effectively throughout the network.

The support from a range of food actors and groups provided the encouragement for OFN to secure grant funding from various community funding agencies to hire an "Ideas in Action" facilitator to develop and grow local projects to become long-lasting, viable activities for people to run in their communities. Demonstrating a united front and collaboration by groups was an important way to highlight to the council how its involvement in food governance could support civil society efforts to transform the food system.

Yet, while framing of food initiatives around local food and community resilience included reference to food access, more deliberate effort was needed to articulate how these broad principles could be operationalized to support the efforts of organizations involved in the charity food sector. For example, while supportive of the efforts of OFN and GFD, leaders in the charity food sector did not see the immediate relevance of these forums to their activities. Rather, they identified accessing funding as their single biggest priority in order to serve an ever-increasing number of residents that were relying on their services to alleviate hunger or improve access to cheaper food. While supportive of broader discussions of food system change, they simply did not have the time or resources to dedicate to those discussions, and as a result attention to food justice and food security were less prominent in the way OFN framed food problems and solutions in Dunedin. The relative lack of attention to food justice and food security concerns applied equally to GFD, while the council position would have unlikely become a reality without the persistent and proactive lobbying pursued by OFN, which relied on the informal and discursive support of civil society that had emerged out of the food forums. Yet once again, voices from the charity food sector were less prominent, which shaped how OFN framed the need for a formal council role as the coalescence of local food and community resilience through economic development. One city councillor recognised that "we have been very lucky, like the Our Food Network, those guys have been awesome in terms of the lobbying, so without those guys, I doubt we would have [the Food Resilience and Business Advisor] position". Following the establishment of the Food Resilience and Business Advisor through the collective effort of a council process and OFN lobbying, one person involved in the process explained that

> there is just huge value in coming together as a group and having those conversations . . . but I guess it's just as much a power in coming together as a group and understanding where everyone is coming from and so yeah we are not necessarily in opposition even though some things aren't going to be perfectly aligned.

The challenge then becomes whose voices are promoted and whose are obscured, as the emergence of direct council engagement in food system change has shifted the governance dynamic, as various individuals and organizations contributed to guiding the work of Good Food Dunedin through the multi-stakeholder Good Food Alliance.

As mentioned above, the GFD Business Advisor position sits under the economic development sector of local government and one of the first jobs for this advisor was to develop a working group, now known as the GFA, to discuss the differing views and agendas around food by stakeholder groups. Early unsuccessful efforts were focused on obtaining a diverse group of stakeholder participation, including large food industry and supermarket representatives. As a council staff member explained "our end goal . . . will be working with everybody whoever they are in that spectrum". However, in reality, participation in GFA meetings primarily includes individuals and groups that are already involved with OFN. The setting and the agenda of GFA meetings are much more focused on the economic opportunities of food system change. For example, programs have been developed to provide advice for food business opportunities, a "one-stop shop" for food business permitting and telling Dunedin's food story for visitors. A representative from the charity food sector made the point that

> I think that's had a couple of benefits I mean [the Food Advisor], even in that team (Economic Development team for the DCC) is shaping the conversation around food, which is really helpful. And what food resilience might look like within an economic development frame.

What we see here is how food governance mechanisms are being developed that bring together the OFN network of individuals and groups that value the fact that food is being discussed with the resources of DCC through GFD to reframe food issues in economic terms in an effort to bring additional people to the table who may not share the same values and motivations.

However, in this process, the diversity of views and approaches is not being discussed. While the open-ended goals of "good food" are inclusive, mechanisms have not yet emerged for critical reflection of food issues or how those issues might be addressed through additional food system governance resources. Deliberate choices to frame food in either "local" or "resilient" terms has minimized the degree to which alternative food actors motivated primarily by issues of food justice see their goals and objectives reflected in either OFN or GFD processes. What was once exclusively a relatively small civil society effort to broaden discussion amongst existing alternative food system actors to ensure that food had a place in the council's decision-making process has resulted in renegotiation and compromise about how food system problems and solutions should be framed and reframed. At the same time this process has dramatically extended the reach and prominence of food system change as an issue for discussion in Dunedin.

Conclusion

What has been experienced in the Dunedin context is an ongoing process of experimentation with regards to how civil society and the council interact around food system governance. The process of engagement has shifted priorities and agendas for all groups, as they are forced to re-evaluate their approaches and projects in the face of a diversifying group of stakeholders that brings a range of views of what they want the Dunedin food system to be. The initial economic focus of GFD used the power of the market as the most effective means of justifying Council involvement. OFN were apprehensive about the overt economic focus, while similarly, other council actors were concerned that without emphasizing the economic development opportunities, they would be unable to justify why discussions about food should be addressed by the council at all. Alternatively, OFN had focused on local food as a means of bringing together groups that coalesced around a desire for a "different" food system. Despite these differences, the place of food governance within DCC and the broader community has been legitimized.

As discussed in Chapter 1, positioning food governance in Dunedin along the governance continuum is a factor of the instrumental resources, organizing frame, and broader political-economic context. The process of food governance in Dunedin is shaped by the uneven access and use of different types of power. OFN and other civil society groups claimed governance space for local food in a vacuum, where formal food policy is exclusively focused on efforts to maximize production for export. However, they did not have the necessary instrumental resources to activate or operate beyond their closed network of already existing food actors. Through efforts to exercise additional instrumental resources and structural power for food system change with direct engagement with the council through GFD, OFN has developed a more pragmatic approach to food system transformation. While council involvement in food governance has unlocked important resources and provided greater legitimacy to food system change efforts, it is not without cost. The increased prominence of economic discourse of food system change can be seen as the price to pay for access to those resources. The emerging process of co-governance in Dunedin has provided opportunities for multiple civil society and local state actors to work together to address local food issues. However, these efforts are constrained by the broader political-economic context of food and agriculture in NZ, where there has been little opportunity to engage and reframe national-level agri-food discussions.

The multifunctional nature of food (Goodman, 2004) presents ongoing challenges in negotiating the diverse and competing claims for what "doing food differently" means for different actors. Yet as Carolan (2013) argues, the process of grappling with that complexity, addressing the inherent tensions and the unsettling of values about and beyond food provides the basis for hopeful imaginaries of food system change. The development of the Dunedin Food Charter is a good example. It outlines an aspirational commitment and an action plan for food that is "good for people, good for the city and good for the planet" (GFD, n.d.b) and provides opportunities for all stakeholders and individuals to relate their specific projects to broader goals of food system change.

The emergence of local food governance in Dunedin has also reshaped civil society and local government relationships, resulting in new networks of support that increase capacity and provide access to new resources to engage in this work. There is an increase in food consciousness, as food is increasingly on the agenda of local decision-making in council planning, through local media and is increasingly visible in the city. However, opportunities to connect mechanisms for food system governance to other ongoing social movements in the city, such as living wage campaigns and ethical procurement, or to connect with food system change beyond place at the regional, nationalm or international scales, have yet to materialize. It remains to be seen if the evolving approach to food system governance in Dunedin can adapt to support different food system outcomes and priorities of a range of actors or if these new governance processes will realize opportunities to carve out new spaces for engagement and contestation about food and beyond.

Acknowledgements

The authors wish to acknowledge the efforts of the many people who are working to ensure that Dunedin's food system becomes more sustainable.

In particular, we would like to thank those individuals who gave up some of their precious time to participate in the research. Thanks also to Chris Garden in the Department of Geography for creating the figures.

References

Amin, S. (2011) Food sovereignty: a struggle for convergence in diversity. In: Holt-Gimenez, E. (ed.) *Food movements unite! Strategies to transform our food systems*. Oakland, Food First Books, pp. i–xviii.

Bartos, A. E. (2014) Discourses of food sovereignty from somewhere. *Dialogues in Human Geography*, 4(2), 190–194.

Blay-Palmer, A., Knezevic, I., & Spring, A. (2014) Seeking common ground for food system transformation. *Dialogues in Human Geography*, 4(2), 185–189.

Booth, S. & Whelan, J. (2014) Hungry for change: the food banking industry in Australia. *British Food Journal*, 116(9), 1392–1404.

Campbell, D. C., Carlisle-Cummins, I., & Feenstra, G. (2013) Community food systems: strengthening the research-to-practice continuum. *Journal of Agriculture, Food Systems and Community Development*, 3(3), 121–138.

Campbell, H. (2005) The rise and rise of Eurep-GAP: European re(invention) of colonial food relations? *International Journal of Sociology of Agriculture and Food*, 13(2), 1–19.

Carolan, M. S. (2013) The wild side of agro-food studies: on co-experimentation, politics, change, and hope. *Sociologia Ruralis*, 53(4), 413–431.

Challies, E. R. T. & Murray, W. E. (2008) Towards post-neoliberalism? The comparative politico-economic transition of New Zealand and Chile. *Asia Pacific Viewpoint*, 49(2), 228–243.

Clapp, J. & Fuchs, D. A. (2009) *Corporate power in global agrifood governance*. Cambridge, MIT Press.

Connelly, S. & Nel, E. (2016) New Zealand: restructuring of the New Zealand economy: global-local links and evidence from the west coast and southland regions. In: Halseth, G. (ed.) *Transformation of resource towns and peripheries*. London, Routledge, pp. 136–160.

Conradson, D. & Pawson, E. (2009) New cultural economies of marginality: revisiting the West Coast, South Island, New Zealand. *Journal of Rural Studies*, 25(1), 77–86.

DCC (Dunedin City Council) (2013) *Community resilience forum recommendations report on food*. Available at: www.dunedin.govt.nz/__data/assets/pdf_file/0020/362612/ma_council_r_food_2013_12_09.pdf

DCC (2015) *Food resilience*. Available at: www.dunedin.govt.nz/__data/assets/pdf_file/0006/500487/Item-17-Food-Resilience.pdf

DCC (n.d.) *Food resilience*. Available at: www.dunedin.govt.nz/your-council/council-documents/energy-plan/food-resilience

DuPuis, E. M. & Goodman, D. (2005) Should we go “home” to eat? Toward a reflexive politics of localism. *Journal of Rural Studies*, 21(3), 359–371.

Dwiartama, A. & Piatti, C. (2016) Assembling local, assembling food security. *Agriculture and Human Values*, 33(1), 153–164.

GFD (Good Food Dunedin) (n.d.a) *About*. Available at: https://www.dunedinnz.com/live-and-work/good-food-dunedin

GFD (n.d.b) *Good food Dunedin charter*. Available at: www.dunedinnz.com/__data/assets/pdf_file/0007/653191/Good-Food-Charter-May-2018.pdf

Goodman, D. (2004) Rural Europe redux? Reflections on alternative agro-food networks and paradigm change. *Sociologia Ruralis*, 44(1), 3–16.

Goodman, D., DuPuis, E. M., & Goodman, M. K. (2012) *Alternative food networks: knowledge, practice, and politics*. London, Routledge.

Guthman, J. (2008) Neoliberalism and the making of food politics in California. *Geoforum*, 39(3), 1171–1183.

Guthman, J. (2011) *Weighing in: obesity, food justice, and the limits of capitalism*. Berkeley, University of California Press.

Hassanein, N. (2003) Practicing food democracy: a pragmatic politics of transformation. *Journal of Rural Studies*, 19(1), 77–86.

Haylock, K. & Connelly, S. (2018) Examining the insider/outsider dimensions of local food system planning: cases from Dunedin and Christchurch, New Zealand. *Planning Practice & Research*. Available at: https://doi.org/10.1080/02697459.2018.1546470

Jarosz, L. (2014) Comparing food security and food sovereignty discourses. *Dialogues in Human Geography*, 4(2), 168–181.

Levkoe, C. Z. (2011) Towards a transformative food politics. *Local Environment*, 16(7), 687–705.

Marsden, T. (2013) From post-productionism to reflexive governance: contested transitions in securing more sustainable food futures. *Journal of Rural Studies*, 29, 123–134.

McClintock, N. (2013) Radical, reformist, and garden-variety neoliberal: coming to terms with urban agriculture's contradictions. *Local Environment*, 19(2), 147–171.

Morris, C. (2018) Council endorses food charter. *Otago Daily Times*, 15 March. Available at: www.odt.co.nz/news/dunedin/dcc/council-endorses-city-food-charter

Morgan, K. (2015) Nourishing the city: the rise of the urban food question in the Global North. *Urban Studies*, 52(8), 1379–1394.

OFN (Our Food Network) (n.d.) *Our Food Network*. Available at: www.ourfoodnetwork.org.nz/

Paci-Green, R. & Berardi, G. (2015) Do global food systems have an Achilles heel? The potential for regional food systems to support resilience in regional disasters. *Journal of Environmental Studies*, 5(4), 686–698.

Peet, J. R. (2012) Comparative policy analysis: neoliberalising New Zealand. *New Zealand Geographer*, 68(3), 151–167.

Sherriff, G. (2008) Towards healthy local food: issues in achieving just sustainability. *Local Environment*, 14(1), 73–92.

Stats NZ (2013) *2013 Census QuickStats about a place: Dunedin City*. Available at: http://archive.stats.govt.nz/Census/2013-census/profile-and-summary-reports/quickstats-about-a-place.aspx?request_value=15022&parent_id=14973&tabname=

4 Catalyzing change in local food system governance in Calgary, Alberta

The role of YYC Growers and Distributors cooperative

Mary Beckie and Elizabeth Bacon

Introduction

Food, as both a product and a practice, impacts us all. Concerns about the cost of food, food quality, food insecurity, environmental degradation, climate change, urbanization, declining numbers of farmers, and increasing corporate control over the food system, are leading to calls for changes throughout the food system. As Pimbert (2012) and others (e.g., Candel, 2014; Hassanein, 2003) argue, the problems and solutions associated with food and agriculture are complex and inherently value-laden, hence the need for more inclusive and collaborative governance involving a broad range of actors and perspectives, appropriate structures, democratic processes, and increased institutional capacity for learning and change. There is, however, no one formula for success. Food system governance and related policy and program development are complicated by the diversity of ecological, political, economic, and cultural processes taking place in or influencing particular regional contexts. Furthermore, the exact composition and roles of those involved, and related relations of power, also impact governance processes and outcomes.

In this chapter we examine the role and influence of a relatively new but rapidly growing social economy organization involved in local food system governance in the city-region of Calgary, Alberta. In contrast to other chapters in this volume that concentrate on the external relationships formed by civil society and social movement actors to advance more collaborative food system governance, this chapter focuses first on the internal, self-governance structure of a producer cooperative and how this has shaped its efforts to collaborate with civil society actors and government in moving towards a more inclusive and co-governed local food system. This organization, YYC Growers and Distributors (YYC), consists of 20 urban and rural growers in and around Calgary that collectively sell their products through a community supported agriculture (CSA) program and at a small number of farmers' markets.[1] Beyond selling fresh produce, YYC is strongly values-driven: committed to environmental and social justice, educating consumers about the value of local food, and influencing policy changes that better support local food systems in a region and a province that has traditionally been dominated by large-scale, export-oriented crop and livestock operations. The conventional agri-food system in Alberta follows a productivist model and is characterized by a heavy reliance on manufactured inputs (ie. large-scale machinery and equipment, agrochemicals) and economies of scale in order to increase the quantity and efficiency of production, and be competitive in the global marketplace (Qualman & Tait, 2004; Lawrence et al, 2001). Small- to medium-scale producers, who are unable or unwilling to follow this dominant model, must devise alternative strategies in order to survive. One of these strategies is the "turn to local".

A surge in consumer demand for locally grown and raised products in Alberta over the past two decades reflects this "turn to local", which has spawned the growth of a number of local marketing options such as farmers' markets, CSA programs, food box schemes, farm-gate retail operations, and local food-oriented restaurants. In 2016, the total market value estimate for farmers' markets, farm retail, and local food restaurants was $1.624 billion in Alberta, almost quadruple the province's total in 2004 (AAF, 2016). While public interest for purchasing local has increased, in part linked to a growing urban and more cosmopolitan population, accompanying policy and structural changes have been slow to emerge. In the City of Calgary, for example, there is no food council in which community, broadly defined, can play an ongoing role in local food governance; however, local food activists and growers have influenced some recent changes to municipal policies and increased support for new programs related to urban agriculture. It is within this context that YYC has caught the attention of policy-makers at both municipal and provincial levels and has played a key role in catalyzing changes in policies and programs. In this chapter, we explore the factors shaping the evolution of YYC, and how the adoption of a cooperative governance structure has enabled the organization to innovate, scale up, build relationships with consumers and community-based organizations, as well as municipal and provincial governments, and thus make a unique contribution to the local food movement in the Calgary region.

This case study utilized a qualitative research approach, with data gathered primarily through semi-structured interviews conducted in Calgary and the surrounding area with growers, board members, customers, and municipal and provincial government representatives. Of the 15 interviews, 6 were conducted via telephone, and 9 were conducted in person, at both urban and rural farms, at a farmers' market, and at a customer's home. In addition to participant observations and informal conversations at a provincial forum on regional food systems (Cultivating Connections, February 2017) and at Calgary's Hillhurst-Sunnyside Farmers' Market (summer 2017), we also referred to websites and web-based resources (YYC, civil society organizations, province of Alberta, City of Calgary), including relevant policy documents and reports.[2]

We begin by first positioning this case study within the literature, focusing on the overlap between the social economy and local food initiatives, food system governance, and the role of agency and relations of power. Next, we provide a brief overview of the provincial and municipal context. In the following sections, we describe the establishment and evolution of YYC, and then examine the self-governance of the cooperative within the broader context of multi-stakeholderism in municipal and provincial local food system governance. In the conclusion we reflect on the significance of alliances and convergences between the social economy, civil society, and the public sector in the effort to create more inclusive and democratically governed food systems.

Convergence of the social economy and the local food movement: Cooperatives and food system governance

The social economy (e.g., non-profit, cooperatives, community organizations, foundations) is playing a prevalent role in Canada's alternative and local food movement (Connelly & Beckie, 2016; Beckie et al, 2012; Connelly et al, 2011).[3] There are numerous examples of the social economy's role in food systems, including, community gardens, food banks, food hubs, farmers' markets, and collective kitchens. These organizations are being formed by citizens working together and taking action to address a range of values and needs not being met by the predominantly neoliberal, globalized food system (Raffaelli, 2016). In contrast to the dominant food system, social economy food initiatives are often locally embedded and controlled (Feagan & Morris, 2009), and are "conditioned by local community norms, values and culture" (Lyson et al, 1995, p. 108). Distinct from private and public sectors, the social economy often interacts with and in some cases partners with these sectors, as well as with members of civil society, in order to achieve common goals. By building relationships that are linked by a common purpose, the social economy aids in creating a space for collaboration, knowledge creation, and innovation that benefits its members and the community as a whole (Fonteneau et al, 2011). Farmer cooperatives that direct-market locally produced food are one such example where collaboration and reciprocity among producers and consumers can achieve mutual goals and benefits that can also extend to the broader community.

In the Canadian prairie region, cooperatives have played an instrumental role in supporting communities and the agricultural sector over the past 100 years (Gertler, 2007; MacPherson, n.d.). For example, farmer cooperatives, such as the Saskatchewan Wheat Pool and the United Farmers of Alberta, were established in the 1920s to provide farmers with a collective infrastructure to get their products to markets and ensure that they received a fair price (Faucher, 1947). Other cooperatives in the Prairie Region provide farmers with access to equipment and financing at fair rates, while others have established electricity and gas distribution in rural communities (Gismondi et al, 2016; Gertler, 2007). These cooperatives, like all cooperatives, are guided by seven key principles sanctioned and promoted by the International Cooperative Alliance (ICA): voluntary and open membership; democratic member control; member economic participation; autonomy and independence; education, training, and information; cooperation among cooperatives; and, concern for community (Zeuli et al, 2004).

On the governance continuum presented in Chapter 1, cooperatives can be characterized as self-governance arrangements, wherein members work together to democratically make decisions and achieve consensus. Producer cooperatives like YYC Growers and Distributors are also embedded in the community, and can form direct relationships with customers, such as through CSA programs or at farmers' markets. The educational opportunities and awareness raising intrinsic in these relationships can, in turn, foster the emergence of "citizen consumers" (Renting et al, 2012, p. 290). Citizens are empowered through their active participation in the food system, having the ability to directly connect to where their food comes from, and becoming more educated about "the limits and possibilities of local food systems" (Beckie et al, 2012, p. 343). Likewise, producers are made more visible through establishing these direct relationships with consumers. By recognizing the social, cultural, and environmental values of food, thereby delivering more than just a commodity or economic transaction, producer cooperatives can reshape their relationship with food systems and hence become "citizen producers" (Renting et al, 2012, p. 290). In this way, they can exert both instrumental power – influence through direct action and use of technical knowledge and experience – and discursive power – the development and acknowledgement of new narratives and norms related to agriculture and food (see Chapter 1). Also, as discussed in the introductory chapter, this agency can influence changes in external political structures and discourses when conditions are favourable, but success in achieving these goals is less likely when conditions are inhospitable.

Interest in more inclusive, democratic, and locally embedded governance models for food systems has been growing in recent decades due to concerns about the negative impacts of the predominantly neoliberal, globalized food system, where transnational corporations are the dominant players and decision-making is based primarily upon economic criteria. However, achieving inclusivity and collaboration in food system governance "takes time, effort and resources" (Rydin & Pennington, 2000, p. 161), and frictions can arise when multiple actors and perspectives are in dialogue. Despite the potential conflicts, these collaborative and dialogical processes "can also create a space for new, more productive patterns to emerge" (Pearce & Littlejohn, 1997, p. 79). Creating these spaces of deliberation for actors to reflect on and discuss diverse views and values of food and governance is referred to as reflexive governance (Hospes & Brons, 2016; Sonnino et al, 2014). In this dynamic and inclusive approach to governing, participants are encouraged to "scrutinize and reconsider their underlying assumptions, institutional arrangements, practices and expectations" (Hendriks & Grin, 2007, p. 333), in order to "learn from each other and ultimately to identify collective solutions to shared problems" (Sonnino et al, 2014, p. 2).

Cities and urban areas, in particular, have become key players in the local food movement and in the governance of local food systems (Hospes & Brons, 2016). This has been influenced by a number of factors, such as unprecedented urbanization (Mansfield & Mendes, 2013), citizen interest in localized food systems (Beckie et al, 2013), demand for supportive food and agriculture policy (Morgan, 2013; Hiley et al, 2011), and the devolution or downloading and offloading of responsibilities from higher levels of government (Hiley et al, 2011). Despite the challenges associated with increasing demands and responsibilities, many municipal governments are beginning to acknowledge the benefits and "multi-functional nature" (Morgan & Sonnino, 2010, p. 210) of urban agri-food systems, such as providing economic opportunities, skill training, social and cultural opportunities, and ecological functions (Beckie & Bogdan, 2010; Mougeot, 2006). Urban agri-food systems are also creating a platform for dialogue, collaboration, and social learning. There are many examples, nationally and internationally, of innovations in food governance structures (e.g. food councils) and targeted municipal and regional policies (e.g., food strategies, land use and zoning bylaws); however the extent of these developments varies significantly from one municipality to the next.

The provincial and municipal context

Alberta is one of Canada's three Prairie provinces situated in the Northern Great Plains Region. A well-developed, resource-based economy has supported Alberta's strong population growth, which is the second-highest rate in Canada and more than double the national average, with 83.6 per cent of the province's population of 4.286 million now living in urban centres (Statistics Canada, 2016). Despite this overall growth, many rural communities are in decline, and from 2006 to 2011, the number of farms and the amount of farmland in Alberta decreased by 12.5 per cent and 3.1 per cent respectively.[4] Much of this farmland consisted of high-quality soils near urban centres. The situation has been exacerbated by fragmentation of farmland due to urban and industrial development and land speculation (Stan & Sanchez-Azofeifa, 2017; Qiu et al, 2015). As a result, farmland prices have been driven beyond agricultural productivity values, making it financially difficult for those farmers wanting to start or expand an operation, while enticing other farmers to sell their land (AARD, 2002). In response, there is growing interest in small-scale farming and urban agriculture as a lower-investment option by those wanting to move into food production for local markets (Statistics Canada, 2016). In 2017, the first provincial forum to bring actors from across the local food spectrum (consumers, community organizations, businesses, government) together, titled Cultivating Connections, had over 80 presenters from across Alberta and drew over 350 people during the two-day event (Alberta Food Matters, 2017). This event was coordinated through a collaboration involving Alberta researchers and community partners associated with FLEdGE (Food: Locally Embedded, Globally Engaged), a cross-Canada research project focused on sustainable regional food systems, and the City of Edmonton.[5]

Alberta's local food sector has received expanded provincial interest and support over the past few years. The province's Department of Agriculture and Forestry, through its Explore Local division, has both funded research and developed a variety of programs associated with food production and marketing. The recent adoption of Bill 7 by the provincial legislature, titled Supporting Alberta's Local Food Sector, partially came into effect June 11, 2018, establishing an Alberta Local Food Week, a provincial Local Food Council, and an organic certification program (Legislative Assembly of Alberta, 2018a). The remainder of the bill, which outlines more specifics of the legislation, will come into effect April 1, 2019 (Legislative Assembly of Alberta, 2018b). Over the next year, the Local Food Policy Council will bring together producers and local food actors from across the province to provide input on what policies, pilot projects, and programs will help boost Alberta's local food sector (Graney, 2018).

The City of Calgary, situated in southern Alberta, has developed a vibrant local food scene, driven in large part by the city's local food restaurants and farmers' markets, which in 2014 had market values of $136 million and $106 million respectively (AARD, 2015). Motivated by this growing demand and public interest in local food, in 2012 the city's food strategy – Calgary Eats! – was developed, framed by the principles of local, accessible, secured supply, environmentally sustainable, healthy, and supportive of community development. In 2017, municipal bylaw changes enabled indoor commercial food production permits (e.g., container growing), but not outdoor commercial urban agriculture. 2017 also saw approval of the establishment of a commercial urban agriculture pilot project to convert eight acres of vacant city land to an urban farm focused on the production and sale of fresh food, with a greenhouse and outdoor growing and research spaces (City of Calgary, 2017). Coordinated and managed by the Compost Council of Canada, with YYC members and other urban farmers providing on-the-ground expertise and advice, the pilot will also be in collaboration with policy-makers and planners at the provincial and municipal levels.

In addition to municipal and provincial government initiatives, various civil society organizations have formed in Calgary that focus on specific aspects of local food systems. For example, in 2013 a small group of concerned citizens started Grow Calgary, a community farm on an 11-acre parcel of land near Calgary's Olympic Park, that donates all food produced to social agencies and food access programs, so that "those vulnerable and living in poverty" can have access to "healthy, local food" (Grow Calgary, 2018). The New Urban Farm Partnership (NUFP), formed in 2013 but now inactive, served to connect urban farmers with landholders and challenge urban agriculture municipal policy barriers. In 2016 NUFP developed the documentary, "Cultivating Calgary's Local Food Resiliency", featuring members of YYC, local non-governmental organizations (NGOs), as well as Calgary municipal councillors, all commenting on the critical role that local agri-food systems currently and potentially play in building food security and urban resilience.[6] In 2017, driven by the dissolution of NUFP, the Calgary Food System Alliance, another non-profit, was formed to cultivate dialogue among local government officials, urban growers, and interested citizens, in order to raise awareness of the value of the local food system and create strategic partnerships to advance it.

In the following section we describe the establishment and evolution of YYC Growers and Distributors, followed by an analysis of their internal governance structure, the relationships of mutual benefit and reciprocity they have developed with their customers, and their connections and influence with civil society organizations and different levels of government. Throughout this section, we use the case of YYC to illustrate the dynamic, complex, and multi-level nature of food system governance, and the ways in which certain actors take agency and influence others to support and join in their efforts of creating a resilient and vibrant local food system. We also discuss the challenges and obstacles faced by YYC and the manoeuvering strategies and compromises they have made.

YYC Growers and Distributors

Establishment and evolution of YYC

YYC was initially incorporated as a not-for-profit society in 2014, the result of a small group of urban farmers in Calgary coming together to pool their produce and resources, and trial a collective CSA. The initial CSA was a success, prompting the growers to form a society, and over the next two years they expanded their customer and grower base, product variety and volume, and physical infrastructure (e.g., transportation, storage). There was recognition of the need to consider a different governance structure as the organization grew and evolved, which resulted in the adoption of a cooperative governance structure in 2017. Each member has equal decision-making power, with a one-member, one-vote policy. Responsibilities of marketing, distribution, storage, and shipping are removed from individual farms and assumed by the cooperative. Up until November 2017, YYC's five-member board of directors oversaw the organization's operations, but a more effective management model has since been adopted, with two experienced members becoming full-time paid managers. YYC has also recently established an external business advisory panel, which has been important for accessing additional expertise in business development.

YYC's overarching mission is "farmers feeding Calgary in markets and homes", with core values including "local, sustainable, soil health, flavour, nutritious, freshness, lower environmental footprint, education, equitable, security, and food democracy" (YYC Growers & Distributors, 2018).

Figure 4.1 YYC's delivery van
Source: Dick Pearson (used with permission)

The cooperative brings together a diversity of products for its CSA, such as: foraged wild mushrooms, micro-greens grown in an urban warehouse, herbs from an urban aquaponics facility, vegetables grown in small urban plots, vegetables grown in a rural greenhouse, as well as large field crop operations. Collectively owned physical infrastructure includes a delivery van and a cold storage warehouse that serves as a central depot and packing site. In addition to two managers and a warehouse employee, YYC has an extensive network of volunteers that assist in various capacities, from farm help, to packing assistance, to promotion and education.

YYC's main focus is their CSA program, though they also sell their products at three farmers' markets in Calgary. Products from growers are combined into a CSA harvest box that is available for pickup by customers in select locations in Calgary. The options include a 25-week CSA starting in May, or the more typical 16-week CSA starting in July. The past two years they also offered a winter box every other week until the end of April. The cost of the CSA is $28 per week for the couple share and $52.50 for the family share. Customers pay for their boxes either in advance or in installations, which allows YYC to provide prepayments to support growers' seasonal start-up costs. In the years that their CSA has been operational, shares have expanded from an initial 75–85 per season, to approximately 650 shares in the 2018 season.

Producer–producer and rural–urban linkages: Contributing to a new agrarianism

The relationships formed among producers within YYC's cooperative structure have enabled knowledge sharing, collaborative decision-making, and collective action, all of which have been crucial to YYC's success and expansion. Collaboration between urban and rural farmers within one organization is unique in Alberta, and contributes to YYC's goal of empowering local farmers to work together to build a more resilient food system. These relationships developed between producers and the pooling and mobilizing of resources have been foundational to the structure, composition, and function of this self-governing cooperative, and its role in Calgary's and Alberta's local food movement. With a membership consisting mostly of young people with limited farming experience, YYC is contributing to changing trends in agriculture in Alberta, as reflected in the increased proportion of new entrants into small-scale agriculture in the 2016 agricultural census (Statistics Canada, 2016). By developing a new narrative on agrarianism, YYC is claiming and enacting discursive power in visible spaces. As discussed in Chapter 1, in order for groups to truly and effectively wield discursive power, they must be considered legitimate within the broader "'field' of social and cultural norms" (p. 27).

The broader field is, in this case, the regional food system: by bringing a diverse group of farmers together, YYC is blurring the boundaries around cities and between urban and rural growers. The organization has created a platform for urban–rural collaboration and innovative problem-solving that meets their needs, values, and priorities. As part of the social economy, YYC members are engaged in "economic activity with a social remit" (Smith, 2005, p. 276), as underscored by their core values of building a sustainable and just food system. In addition to members working collaboratively towards that goal, YYC also demonstrates the social economy principles of inclusiveness and reciprocity. This was exemplified during a mixer event organized in 2017 by the Explore Local division of Alberta Agriculture and Forestry, where urban and rural farmers from the Calgary region, including those of YYC, were invited to meet and network. According to the government employee who coordinated the event, YYC's presence was essential to the event's success; growers were open and willing to share technical knowledge with other producers and new entrants, such as lessons they have learned from being part of a new and rapidly growing cooperative. The technical and marketing expertise that YYC is accumulating is also generating instrumental power that is being noted by others in the industry as well as by municipal and provincial governments.

As a self-governed and democratic organization, all voices and perspectives are heard and contribute equally to YYC's decision-making process. There are regular board meetings throughout the year, as well as annual general meetings in the winter where members review the past season's performance and make decisions about the upcoming growing season and the overall trajectory for the organization. It was through this reflexive governance process that the decision was made to transition from a not-for-profit society to a formal cooperative, and from having board members share responsibilities for daily business operations to hiring two members as managers. This demonstrates YYC's continuous assessment and reconsideration of its organizational structures and assumptions, along with collaborative learning, decision-making, and action (Sonnino et al, 2014). According to members interviewed, becoming a formal cooperative involved a significant learning curve and a considerable amount of work, but formalization has reinforced YYC's democratic values and principles. Becoming a cooperative has also provided more business legitimacy, which has helped to raise the profile of the organization within the food industry and government.

Producer–customer linkages: YYC's role in local food system education and awareness

Beyond achieving its business goals, YYC aims to connect with consumers to raise awareness of the potential of local food systems to achieve social, economic, and environmental goals. YYC aims to reconnect people to the value of land for food production, as opposed to viewing land as only for residential or industrial development. As one participant in the previously discussed NUFP documentary noted, "urban agriculture makes farming, growing food, the labour involved, and the products produced visible". YYC's CSA and presence at farmers' markets are opportunities for customers to engage with producers and learn more about their food and how it is produced. Many growers noted that a key benefit of their membership in the cooperative was sharing information with customers and developing relationships with them. The theme of reconnection is further reinforced by YYC's home pick-up program. This arrangement provides customers the opportunity to meet each other while picking up their weekly CSA box, which one of YYC's general managers views as "the most beautiful, perfect amalgamation of things that we should be doing, in terms of community economic development, grassroots connecting people with the people, and empowering their foods". A customer, whose home is the pick-up point for her neighbourhood, said that this "enriches the community as well. It's just nice to be able to meet the neighbours and have some relationships with people that you may not have had before". The homeowner also described how she acted as a point of knowledge exchange for her neighbours, giving them information about the produce in the weekly share. This illustrates the development of co-governance relationships, wherein consumers can also take on an active role, becoming "citizen-consumers . . . regain[ing] control over the ways in which their food is produced and provided" (Renting et al, 2012, p. 290).

Figure 4.2 YYC's stall at a Calgary farmers' market
Source: Dick Pearson (used with permission)

Knowledge exchange also occurs through YYC's social media and web pages. Facebook is a key channel for informing customers and raising the organization's profile, particularly with the "150 Faces of YYC Growers" campaign, where YYC released 150 photographs of individuals associated with the organization, accompanied by a short caption describing the individual's connection to YYC or the broader Calgary food system. In conjunction with Canada 150 celebrations in 2017, this campaign generated significant traffic to YYC's Facebook page and is a case of raising awareness and educating consumers about the regional food system. These activities are viewed as crucial by YYC members, given the general lack of understanding among the public and in government about the importance and real cost of locally grown food; as one member commented:

> People don't understand what food tastes like, they don't understand that paying more for food that is more nutritious and flavourful makes more sense. People don't understand that you can grow food here at the level you can; governments don't understand it either because they're in the same mindset and they don't see [urban] farming as gainful employment.

Given YYC's labour requirements and inability to externalize their costs and risks, like large-scale agri-food businesses, produce prices are typically higher than what appears in grocery stores, which makes purchasing food from the cooperative unaffordable for some citizens.[7] As one grower explained: "I have to charge enough for my products [to be viable], but this puts it out of reach for some Calgarians; the fiscal reality is a real challenge". To address this situation, YYC donates fresh produce to the Alex Community Food Centre, the Mustard Seed, and other social service agencies that are not only providing access to local food for those in need, but are also using food as a way to build individual and community capacity and resilience.

YYC's relationship with civil society organizations and government

In order to understand YYC's development and broader influence on the regional food system, it is important to consider not only the relationships formed between growers and with consumers, but also YYC's relationship with civil society organizations and government. With a democratic cooperative structure that is connected to other local food actors, YYC is playing a pivotal role in the development of a "civic food network" (Anderson et al, 2014). As noted earlier, a goal of the social economy is to build relationships that are linked by a common purpose, creating spaces for collaboration among the different actors who enact specific roles in the network (Fonteneau et al, 2011). YYC's direct action on-the-ground makes visible the physical and social components of an urban food system; civil society organizations educate and advocate with and on behalf of local producers for a more sustainable food system; and, government's role is to respond with changes in policies and programs. But implicit in these relationships, as in all relationships, is the dynamic and continuous exercise of power and how those relationships of power affect governance.

On the governance continuum described in Chapter 1, YYC operates in the realm between multi-stakeholderism, where engagement of food movement actors is weak, and polycentric governance, where engagement is strong and food movement actors are part of the governing structure. In examining and applying Gaventa's power cube (2005, 2007), YYC can be characterized as having claimed and created spaces in Calgary's emerging local food movement, visibly exercising instrumental and discursive power. When governance and power are considered within municipal and provincial political contexts, however, the agency of local food actors, like YYC, is reduced within the predominantly multi-stakeholder approach utilized by the city and the province. YYC and other local food actors are consulted on decisions by government, but they do not have equal power in the final decision-making processes, where power is more obscured due to the influence of other actors, such as land investors and developers.

The shift towards co-governance or polycentric governance from multi-stakeholder governance regimes is politically and economically challenging for local governments. In Canada, many municipalities lack the authority, resources, and political will to make some of the bigger, more transformative changes needed to create more sustainable and resilient food systems, such as the protection of farmland (Hiley et al, 2011). Developing more supportive policies to protect land for agriculture conflicts with the pressure for growth through residential and industrial development. Cities are reliant on property taxes for most of their financial base and can generate greater revenue from non-agricultural property taxes or from selling public land to developers rather than through maintaining it for urban agricultural use (Beckie et al, 2013). Urban municipalities' contribution to the loss of productive farmland has been particularly evident in Alberta, where most of the high-quality farmland is located between the rapidly growing cities of Calgary and Edmonton (Beckie et al, 2013). As one YYC member commented, "urban sprawl is bleeding good land".

Many local food advocates in Calgary feel frustrated that recognition and support for a diverse and integrated local food system by government has been limited and slow to develop. As one grower noted, instead of creating effective policies to enable and empower small agri-food businesses, governments look at urban growers as "a nuisance thing to deal with", despite their contribution to a rapidly growing local food sector. Without being eligible to obtain a business license or owning most of the land they farm, and often earning less than $10,000 per year, urban farmers cannot qualify for provincial farming subsidies (such as fuel rebates) or for federal agriculture business development grants. Despite these obstacles, YYC continues to advocate for more transformative changes by strategically building alliances with civil society organizations, their customers, and with allies within local and provincial government. Recent changes in municipal and provincial policies and structures (e.g., Calgary's commercial urban farm pilot project; Alberta's Local Food Act and provincial Local Food Council) suggest that these efforts are having a positive impact in what appears to be a more receptive socio-political climate. According to a representative of the Explore Local division of the Department of Agriculture and Forestry, the provincial government is now taking local food more seriously due to a rapidly expanding local food economy, along with the lobbying efforts of civil society and local food businesses like YYC. With a provincial election on the horizon, however, it is difficult to say whether these changes will lead to long-lasting transformation of local food system governance in Alberta.

Conclusion

Given the complexity and place-based nature of local food systems, there is clearly no one path to the development of more inclusive and democratic governance models. Most of the food governance literature focuses on what governance should or ought to aspire to, instead of how governance is currently functioning in practice (Candel, 2014). While the normative principles of democratic and reflexive governance (i.e. dialogue, critical analysis, social learning, innovation, and capacity for change) are fundamentally important, Candel (2014) calls for more empirical research investigating current agri-food governance arrangements and their challenges and successes.

In this case study we examined the nature of regional food system governance in Calgary, Alberta, and the influential role that YYC Growers and Distributors, a local food producer cooperative, is playing. As a social economy organization that is building knowledge, expertise, and collective agency, YYC is wielding both instrumental and discursive power. It is part of a young agrarian movement taking place in Alberta and across the country, and is breaking down barriers and strengthening ties between the urban and the rural. In this way, YYC is also contributing to a "new localism", a "refashioned foodscape in which the city, the countryside and all different actors and stakeholders that occupy their spaces are reconnected physically, culturally, environmentally, socially and economically" (Sonnino, 2017, p.4). YYC has been built upon values that are embedded in the cooperative and reflexive self-governance structure that aims to connect growers in a dynamic and responsive way. But in alignment with the social economy's ethos of inclusiveness and solidarity, YYC also establishes relationships with other individuals and organizations within the local food movement, in order to create greater agency and momentum for change in local food governance. Operating at the boundary between urban and rural, and participating in different types and levels of governance, YYC is serving as a catalytic and integrative agent.

At different times in the regional food system, YYC has occupied or interacted with all potential spaces in Gaventa's power cube (2005, 2007): closed spaces, created by the barriers to their business legitimacy and development; invited spaces, experienced through their participation and consultation in various municipal and provincial fora; and claimed/created spaces, developed as a result of their different strategies and actions to establish and grow their business, and their relationships with other local food actors. In interacting in these various spaces, and confronting both hidden and visible power, YYC has experienced success, but they have also faced a range of challenges as they "navigate the difficult terrain of remaining viable alternatives to the conventional food system within their local contexts, while also contributing to a broader social movement that uses food as a platform and a catalyst for change" (Connelly & Beckie, 2016, p. 52). The reflexive nature of YYC's internal governance and their ability to manoeuvre, adapt, and innovate within changing political and economic circumstances have been crucial to the organization's viability and to their role in the local food movement. Their vision for a more innovative and resilient food system, and the role of urban agriculture in particular, is described by one YYC member in the 2016 NUFP documentary:

> With the declining number of farmers, we're going to need new people innovating and creating a culture around food. Why not make urban agriculture a legitimate occupation? Each neighbourhood could have their own farmers, integrating into the life of the city. They could build relationships with other farmers – sharing knowledge from one space and region to the next . . . A resilient food system in Calgary is always going to be a complex web of many parts – urban agriculture is part of the web. YYC is a group of young pioneers in Calgary who have made agrarian urbanism happen. We're on the cusp of major change, as food security is an issue for all of us.

By operating at the intersection of the social economy and local food systems, YYC offers a physical and social infrastructure for urban and rural producers, and for citizens interested in learning about, purchasing, and advocating for local food. Producer cooperatives, like YYC, thus have the potential to build and mobilize social capital (Tregear & Cooper, 2016), by enabling individual and organizational collaborators "to become involved in advocacy and influence policy agendas by which they are affected" (Fonte & Cucco, 2017, p. 293). The alliances and convergences formed between social economy organizations and the local food movement relate to their common characteristics, values, and goals: locally embedded, locally controlled, and addressing local needs through relationship building, collaboration, and innovation. The inclusive and reflexive governance orientation of cooperatives, and of the social economy in general, can provide insights for more progressive changes in food system governance. According to Fonte and Cucco (2017), "this process of alliances and convergences between the cooperative movement and the food movement could become an important step in the struggle for sustainable transformation of the food system" (p. 292). Bringing diverse actors together and achieving more inclusive and co-governed food systems is a challenging and ongoing process that requires "appropriate structures, capacity and political will" (Candel, 2014, p. 593). Harnessing the values, commitment, and capacity of social economy organizations can aid in generating innovation and change.

Acknowledgements

We would like to acknowledge the Social Sciences and Humanities Council of Canada (SSHRC) for supporting this research through the FLEdGE Partnership Grant, and the University of Alberta's Faculty of Extension for additional funding. Our sincerest gratitude is also extended to Kye Kocher for input on this chapter, and to the other members of YYC Growers and Distributors who contributed their time and insights to this research.

Notes

1 According to the United States Department of Agriculture (1993), CSAs can be defined as follows:

> In basic terms, CSA consists of a community of individuals who pledge support to a farm operation so that the farmland becomes, either legally or spiritually, the community's farm, with the growers and consumers providing mutual support and sharing the risks and benefits of food production. Typically, members or "share-holders" of the farm or garden pledge in advance to cover the anticipated costs of the farm operation and farmer's salary. In return, they receive shares in the farm's bounty throughout the growing season, as well as satisfaction gained from reconnecting to the land and participating directly in food production. Members also share in the risks of farming, including poor harvests due to unfavorable weather or pests. By direct sales to community members, who have provided the farmer with working capital in advance, growers receive better prices for their crops, gain some financial security, and are relieved of much of the burden of marketing.

2 The sources are as follows: AARD (2015); Alberta Food Matters (2017); Calgary Food System Alliance (2018); City of Calgary (2012); Grow Calgary (2018); Alex Community Food Centre (2018); YYC Growers and Distributors (2018).

3 The following definition is adapted from the International Encyclopedia of the Social Sciences (ww.encyclopedia.com/social-sciences/applied-and-social-sciences-magazines/social-economy):

> The social economy is generally acknowledged as the third sector of mixed capitalist economies, distinct from the private and public sectors. It is based on cooperative, not-for-profit, and voluntary activities carried out within communities, across national economies, and internationally. Organizations within the social economy are classified as cooperatives, mutual societies, voluntary organizations, foundations, and social enterprises; participation and membership is voluntary. Many social economy organizations simply deliver services to their members or others; other social economy organizations, known as social enterprises, engage in trade activities in order to benefit their members or those they serve. In this latter case, any surpluses or profits earned are reinvested in the enterprise, distributed to stakeholder groups, or used for the benefit of those served by the enterprise. Governance typically operates through the "one member, one vote" principle or through enterprise trustees. Recognition of the importance of social economy enterprises is relatively recent, but they are now seen to play significant roles in addressing needs unmet by the private and public sectors.

4 Unlike British Columbia, Ontario, and Quebec, Alberta has no legislation protecting farmland.

5 FLEdGE is a research and knowledge sharing partnership that is committed to fostering food systems that are socially just, ecologically regenerative, economically localized, and that engage citizens. More information can be found at the following website: https://fledgeresearch.ca/

6 See New Urban Farm Partnership (2016).

7 Calgary is a city of extreme wealth and poverty, with income inequality being four times higher than the national average (CPAC, 2017).

References

AAF (Alberta Agriculture and Forestry) (2016) *Study of local food demand in Alberta 2016: consumer study report*. Available at: www1.agric.gov.ab.ca/$Department/deptdocs.nsf/all/explore16484/$FILE/SurveyReport.pdf

AARD (Alberta Agriculture and Rural Development) (2002) *Loss and fragmentation of farmland*. Available at: www1/agric/gov/ab/ca/$department/deptdocs.nsf/all/psc4786

AARD (2015) *Report on local food demand in the Calgary region*. Available at: www1.agric.gov.ab.ca/$Department/deptdocs.nsf/all/explore16031/$FILE/LocalFoodDemand-CalgaryFINAL.pdf

Alberta Food Matters (2017) *Cultivating connections 2017*. Available at: http://www.foodsecurityalberta.org/cultivating-connections-2017-0

Alex Community Food Centre (The) (2018) *The Alex Community Food Centre home page*. Available at: http://thealexcfc.ca/

Anderson, C. R., McDonald, W., Gardiner, J., & McLachlan, S. (2014) Navigating the fault lines in civic food networks. *Journal of Agriculture, Food Systems and Community Development*, 4(3), 79–99.

Beckie, M. & Bogdan, E. (2010) Planting roots: urban agriculture for senior immigrants. *Journal of Agriculture, Food Systems and Community Development*, 1(2), 77–89.

Beckie, M. & Connelly, S. (2016) The role of the social economy in scaling up alternative food initiatives. In: Gismondi, M., Connelly, S., Beckie, M., Markey, S., & Roseland, M. (eds) *Scaling up: the convergence of social economy and sustainability*. Edmonton, Athabasca University Press, pp. 59–81

Beckie, M. A., Kennedy, E. H., & Wittman, H. (2012) Scaling up alternative food networks: farmers' markets and the role of clustering in western Canada. *Agriculture and Human Values*, 29(3), 333–345.

Beckie, M A., Hanson, L., & Schrader, D. (2013) Farms or freeways? Citizen engagement and municipal governance in Edmonton's food and agriculture strategy development. *Journal of Agriculture, Food Systems, and Community Development*, 4(1), 15–31.

Calgary Food System Alliance (2018) *Calgary Food System Alliance home page*. Available at: www.yycfoodalliance.ca/

Candel, J. J. L. (2014) Food security governance: a systematic literature review. *Food Security*, 6(4), 585–601.

City of Calgary (2012) *Calgary eats! A food system assessment and action plan for Calgary*. Available at: www.calgary.ca/CA/cmo/Documents/CalgaryEATS!%20SUMMARY%20Food%20System%20Assessment%20%20Action%20Plan%20for%20Calgary%20May2012.pdf?noredirect=1

City of Calgary (2017) *Urban commercial agriculture pilot project*. Available at: www.calgary.ca/CA/cmo/Pages/Urban-Commercial-Agriculture-Pilot-Project.aspx

Connelly, S. & Beckie, M. (2016) The dilemma of scaling up local food initiatives: is social infrastructure the essential ingredient? *Canadian Food Studies*, 3(2), 49–69.

Connelly, S., Markey, S., & Roseland, M. (2011) Bridging sustainability and the social economy: achieving community transformation through local food initiatives. *Critical Social Policy*, 31(2), 308–324.

CPAC (Chartered Professional Accountants of Canada) (2017) *Income inequality in Canada: the urban gap*. Available at: www.cpacanada.ca/en/the-cpa-profession/about-cpa-canada/key-activities/public-policy-government-relations/economic-policy-research/income-inequality-canada

Faucher, A. (1947) Cooperative trends in Canada. *Annals of the American Academy of Political and Social Sciences*, 253, 184–189.

Feagan, R. & Morris, D. (2009) Consumer quest for embeddedness: a case study of the Brantford Farmers' Market. *International Journal of Consumer Studies*, 33(3), 235–243.

Fonte, M. & Cucco, I. (2017) Cooperatives and food networks in Italy: the long road towards a social economy in agriculture. *Journal of Rural Studies*, 53(1), 291–302.

Fonteneau, B., Neamtan, N., Wanyama, F., Pereira Morais, L., & de Poorter, M. (2011) *Social and solidarity economy: building a common understanding*. Montreal, Canada, Social and Solidarity Academy.

Gaventa, J. (2005) *Reflections on uses of the "power cube" approach for analyzing the spaces, places, and dynamics of civil society participation and engagement*. Brighton, Institute of Development Studies.

Gaventa, J. (2007) Levels, spaces and forms of power: analysing opportunities for change. In: Berenskoetter F. & Williams, M. J. (eds) *Power in World Politics*. New York, Routledge, pp. 204–224.

Gertler, M. (2007) Production co-operatives and co-operative production: legacy and promise. In: Knuttila, M. & Stirling, B. (eds) *The prairie agrarian movement revisited: centenary symposium on the foundation of the territorial grain growers association*. Regina, Canadian Plains Research Center, pp. 107–126.

Gismondi, M., Connelly, S., Beckie, M., Markey, S., & Roseland, M. (eds) (2016) *Scaling up: the convergence of social economy and sustainability*. Athabasca, Athabasca University Press.

Graney, E. (2018) Alberta organics, local food week gives boost to producers, *Edmonton Journal*, 5 April. Available at: http://edmontonjournal.com/news/politics/alberta-organics-local-food-week-gives-boost-to-producers

Grow Calgary (2018) *Grow Calgary home page*. Available at: www.growcalgary.ca/

Hassanein, N. (2003) Practicing food democracy: a pragmatic politics of transformation. *Journal of Rural Studies*, 19(1), 77–86.

Hendriks, C. M. & Grin, J. (2007) Contextualizing reflexive governance: the politics of Dutch transitions to sustainability. *Journal of Environmental Policy & Planning*, 9(3–4), 333–350.

Hiley, J., Bonneau, G., Thomas, K., & Rousseau, M. (2011) Canadian municipalities satisfy craving for sustainable agriculture and food systems. *Municipal World*, 121(4), 27–30.

Hospes, O. & Brons, A. (2016) Food system governance: a systematic literature review. In: Kennedy, A. & Liljeblad, J. (eds) *Food systems governance: challenges for justice, equality, and human rights*. London, Routledge, pp. 20–45.

Lawrence, G., Knuttila, M., & Gray, I. (2001) Globalization, noe-liberalism, and rural decline. In: Epp, R. and Whitson, D. (eds) *Writing off the rural west: globalization, governments and the transformation of rural communities*. Edmonton, University of Alberta Press, pp. 89–108.

Legislative Assembly of Alberta (2018a) *Bill status report for the 29th legislature – 4th session*. Available at: www.assembly.ab.ca/pro/bills/20180308_1200_01_bsr.pdf.

Legislative Assembly of Alberta (2018b) *Bill 7: supporting Alberta's local food sector act*. Available at: www.assembly.ab.ca/ISYS/LADDAR_files/docs/bills/bill/legislature_29/session_4/20180308_bill-007.pdf

Lyson, T. A., Gillespie, G. W., & Hilchey, D. (1995) Farmers' markets and the local community: bridging the formal and informal economy. *American Journal of Alternative Agriculture*, 10(3), 10–113.

MacPherson, I. (n.d.) *The history of the Canadian co-operative movement: a summary, a little historiography, and some issues*. Available at: http://socialeconomyhub.ca/sites/socialeconomyhub.ca/files/Cdian%20Coop%20History.doc

Mansfield, B. & Mendes, W. (2013) Municipal food strategies and integrated approaches to urban agriculture: exploring three cases from the global north. *International Planning Studies*, 18(1), 37–60.

Morgan, K. (2013) The rise of urban food planning. *International Planning Studies*, 18(1), 1–4.

Morgan, K. & Sonnino, R. (2010) The urban foodscape: world cities and the new food equation. *Cambridge Journal of Regions, Economy and Society*, 3(2), 209–224.

Mougeot, L. J. A. (2006) *Growing better cities: urban agriculture for sustainable development*. Ottawa, International Development Research Centre (IDRC).

New Urban Farm Partnership (2016) *Cultivating Calgary's local food resiliency* [video]. Available at: https://vimeo.com/160987626

Pearce, W. & Littlejohn, S. (1997) *Moral conflict: when social worlds collide*. Thousand Oaks, Sage.

Pimbert, M. (2012) *Putting citizens at the heart of food system governance*. London, International Institute for Environment and Development (IIED).

Qiu, F., Laliberté, L., Swallow, B., & Jeffrey, S. (2015) Impacts of fragmentation and neighbor influences on farmland conversion: a case study of the Edmonton-Calgary Corridor, Canada. *Land Use Policy*, 48, 482–494.

Qualman, D. & Tait, D. (2004) *The farm crisis, bigger farms, and the myths of competition and efficiency*. Ottawa, The Canadian Centre for Policy Alternatives.

Raffaelli, P. (2016) Social and solidarity economy in a neoliberal context: transformative or palliative? The case of an Argentinean worker cooperative. *Journal of Entrepreneurial and Organizational Diversity*, 5(1), 33–53.

Renting, H., Schermer, M., & Rossi, A. (2012) Building food democracy: exploring civic food networks and newly emerging forms of food citizenship. *International Journal of Sociology of Agriculture and Food*, 19(3), 289–307.

Rydin, Y. & Pennington M. (2000) Public participation and local environmental planning: the collective action problem and the potential of social capital. *Local Environment*, 5(2), 153–169.

Sonnino, R. (2017) The cultural dynamics of urban food governance. *City, Culture and Society*. Doi: 10.1016/j.ccs.2017.11.001.

Sonnino, R., Torres, C. L., & Schneider, S. (2014) Reflexive governance for food security: the example of school feeding in Brazil. *Journal of Rural Studies*, 36, 1–12.

Smith, G. (2005) Green citizenship and the social economy. *Environmental Politics*, 14(2), 273–289.

Stan, K. D. & Sanchez-Azofeifa, A. (2017) The Edmonton–Calgary corridor: simulating future land cover change under potential government intervention. *Land Use Policy*, 63, 356–368.

Statistics Canada (2016) *Data products, 2016 census*. Available at: www12.statcan.gc.ca/census-recensement/2016/dp-pd/index-eng.cfm

Tregear, A. & Cooper, S. (2016) Embeddedness, social capital and learning in rural areas: the case of producer cooperatives. *Journal of Rural Studies*, 44(1), 101–110.

USDA (US Department of Agriculture) (1993) Defining community supported agriculture. Excerpt from *Community supported agriculture (CSA): an annotated bibliography and resource guide*. USDA & Alternative Farming Systems Information Center (AFSIC).

YYC Growers and Distributors (2018) *YYC Growers and Distributors home page*. Available at: www.yycgrowers.com/

Zeuli, K. A., Cropp, R., & Schaars, M. A. (2004) *Cooperatives: principles and practices in the 21st century*. National Cooperative Business Association. Available at: http://socialeconomyaz.org/wp-content/uploads/2011/06/Zeuli-Cropp-Cooperatives1.pdf

5 Policy engagement as prefiguration

Experiments in food policy governance through the national food policy dialogue in Canada

Charles Z. Levkoe and Amanda Wilson

Introduction

Over the past decade, studies have documented the successes and limitations of place-based initiatives that aim to promote healthy, equitable, and sustainable food systems (Alkon & Guthman, 2017; Levkoe, 2011; Wittman et al, 2011). This research also investigates the ways that these initiatives have become part of broad-based networks that connect a range of actors across sectors, scales, and places (Constance et al, 2014; Holt-Gimenez, 2011; Levkoe & Wakefield, 2014). For Canadian food movements, food sovereignty has become a master frame that addresses the political underpinnings of the dominant food system based on the grounded knowledge of farmers, Indigenous peoples, pastoralists, and fisherfolk. Food sovereignty is "the right of peoples to healthy and culturally appropriate food produced through ecologically sound and sustainable methods and their right to define their own food and agricultural systems" (Nyéléni, 2007). It emphasizes the democratization of food systems, policy, practice, knowledge, and the rights and autonomy of food producers (Wittman et al, 2010; Patel, 2009). Despite a growing interest in broad-based food networks, little attention has been given to how social movements engage with the state to scale-up experiences and learnings from local projects to impact food-related policy while also maintaining goals of social, ecological, and economic justice.

This chapter explores the efforts of social movement organizations[1] to promote empowerment and food systems transformation through engagement in government-led policy-making processes. Specifically, we ask how social movements can advance food policy, while also modeling alternative food futures through processes of policy engagement. We pay particular attention to the ways that different aims coexist, teasing out the tensions, possibilities, and the overall complexity of their interactions. To illustrate these opportunities, we draw on a case study that explores the engagement of a diversity of social movement organizations in the development of a Food Policy for Canada between May and September 2017.[2] Initiated by the federal government, the consultation period included the participation of a wide range of different stakeholders with an aim to establish a national vision for the health, environmental, social, and economic goals related to food.[3]

Both in response, and in parallel to this policy consultation process, Food Secure Canada (FSC), a pan-Canadian food movement alliance, led a series of activities involving more than 70 member organizations and allies in an attempt to prefigure a food systems approach to policy while modeling participatory food governance and strengthening the capacity of Canadian food movements. However, despite meaningful attempts to prefigure more inclusive and equitable food systems through these policy-making processes, these kinds of engagements nonetheless risk limiting a more radical and visionary politics by pressuring social movement actors to prioritize language and approaches that fit within pre-existing government frameworks and that are most easily translated into policy.

Reflecting on the intersections of food systems governance and civil society mobilization, we argue that social movement networks have an ability to prefigure collaborative processes of engagement to advance policy change while strengthening social relationships, deepening their analysis, and advancing collective strategies for change. Our research is based on the authors' active participation in these movements as part of a community–academic partnership through the Community First: Impacts of Community Engagement (CFICE) project.[4] Using primary document analysis, participant observation, and reflecting on our personal experiences in this work, we use the concept of prefiguration (i.e. enacting the desired future in the present) to explore how food movement organizations negotiate and maneuver within the complex terrain of government-led policy-building, simultaneously being grounded in current realities while working to enact alternative food futures.

Our case study represents a point of tension on the governance engagement continuum that frames the analysis of this book. While the social movement organizations that engaged with the national food policy process expressed a desire to establish a more inclusive policy engagement process there was a constant pull towards multi-stakeholderism and an emphasis on consultation rather than meaningful participation. In other words, while FSC and its allies attempted to prefigure grassroots, democratic processes that embedded social and ecological justice into policy, this was complicated by limited resources, conflicting ideals, and power differentials. Participants were under no illusions that a government-led food policy-making process would be transformed into a tool to achieve food sovereignty, rather they saw the policy-building process as a strategic opportunity to enact both instrumental and discursive power to address both short- and long-term challenges in Canada's food system. At the same time, actors had different orientations to these policy-building and engagement activities, which caused some fissures within the movements.

We begin by outlining the context of Canada's food systems and the federal government's consultation for a national food policy, followed by an overview of food systems governance and prefiguration in the policy process. We then present our case study though an account of food movements' engagement with the national food policy consultations. In our discussion, we address the opportunities and limitations for prefiguring a participatory and integrated policy by reflecting on the government's consultation process.

Food policy and Canada's food system

Food policy can be described as a collection of decisions – including laws, regulations, rules and guidelines – that influence the human and non-human relationships across the food chain (Lang et al, 2009). These policies result in a web of complex interactions that shape the broader context in which food-related activities take place, creating an enabling or constraining environment for different governing values and principles. While typically thought of as the sole purview of the state, the realm of food policy includes a range of state and non-state actors, a point we return to below. In this chapter, we focus on Canada's national food policy consultations, representing a government-led policy-making process. In May 2017, Canada's federal minister of agriculture and agri-food launched consultations to inform the establishment of a national food policy. The announcement followed a commitment in the minister's mandate letter to "develop a food policy that promotes healthy living and safe food by putting more healthy, high-quality food, produced by Canadian ranchers and farmers, on the tables of families across the country".[5]

This move was seen as a significant achievement in a long campaign by food movements advocating for a more integrated and "joined-up" approach to food system governance in Canada. Food movement practitioners and scholars have argued that the approach to food policy in Canada has not only been ineffective in tackling food system challenges, but in many cases, it has contributed to additional problems. These include alarming rates of food insecurity among Indigenous and northern communities (Statistics Canada, 2014; Power, 2008), high levels of diet-related disease (Institute for Health Metrics and Evaluation, 2010), dwindling numbers of small-scale farmers and a decrease in farm land (Statistics Canada, 2017; Beaulieu, 2015), costly and unnecessary amounts of food waste (Gooch & Felfel, 2014), unsustainable levels of greenhouse gas emissions attributable to the food system (Environment Canada, 2014), and declining biodiversity of crop, farmed animal, and wildlife species (IPES-Food, 2016) – to name only a few.

The existing patchwork of programs and policies separates food-related sectors and issues, while limiting the effectiveness and scope of existing policy and regulatory interventions (MacRae & Winfield, 2016; MacRae, 2011). As a result, the food policy landscape lacks common principles and priorities to connect jurisdictions and departments that impact the food system. For example, the mandate of Health Canada to maintain and improve the health of all Canadians is often undermined by Agriculture and Agri-Food Canada's directive towards economic growth, which primarily involves large-scale industrial agriculture and export-oriented production. This lack of coordination and collaboration works to divide and isolate pressing food-related challenges, many of which have common structural causes. As a result, advocates argue that most solutions tend to be symptom oriented, incapable of tackling the underlying structural causes (Levkoe, 2011). In response, advocates have called for a more integrated food policy-making process that utilizes a systems approach (Candel & Pereira, 2017) to increase collaboration across government jurisdictions and departments as well as across societal sectors.

These concerns can be thought of as "wicked problems", which refers to addressing a complex set of policy challenges that cut across a wide range of sectors, scales, and jurisdictions (Rittel & Webber, 1973). Responses to these kinds of challenges must go beyond only addressing the symptoms, which is challenging because solutions often become points of contention where stakeholders do not necessarily agree on the problems or the causes. Further, wicked problems have no universal solution and can look different depending on the specific context. Structural solutions must address ways that Canada's food system can feed the population while addressing health, environment, and social equity. Solutions that address the symptoms, rather than the causes, can create additional tensions. For instance, most agree that the solution to ending food insecurity lies in raising income levels, yet we continue to direct resources towards stop-gap measures that fail to address this underlying issue.

A national food policy has been identified as one important way to tackle these wicked problems, the various contradictions, and persistent challenges across the food system, from a systems perspective. In particular, food movements have argued that a national food policy framework rooted in food sovereignty could inform and guide a comprehensive set of policies and programs to ensure consistency and continuity in the management of Canada's food system (FSC, 2017a; PFP, 2011; Wittman et al, 2011). This approach necessitates meaningful engagement with diverse civil society actors not traditionally included in policy-making circles (Andrée et al, 2011; MacRae, 2011; Barling et al, 2002).

Canadian food movements and governance

The development of a Food Policy for Canada is taking place in a context where governance spaces have changed dramatically over the past decades. As outlined in the introduction, since the 1970s, there has been a shift from government (coordination through hierarchy) to governance (coordination through networks) involving new and different technologies of power (Isin, 2000). These networks include state institutions as well as non-state organizations, including private and third sector actors such as non-governmental organizations (NGOs) and voluntary organizations (Mitchell, 2001). The goal of reducing the state role has been pursued by international institutions (e.g. World Bank, International Monetary Fund) through the promotion of "good governance" and "accountability" (Jessop, 2002). While reducing the power of the public sphere and putting increased faith in the free market as an implicit governing principle, these shifts have also opened new spaces for non-state actors to interact with government at various levels that were previously impenetrable (Ilcan & Basok, 2004). The Canadian federal government appears to have embraced this evolution to a more open and transparent approach to policy development, holding numerous consultations across the country with civil society actors and stakeholders on a wide variety of policies and programs. However, as our research findings show, the invitation to consult does not necessarily lead to meaningful engagement in policy-building. In addition, McKeon (2017) warns that participation of diverse stakeholders can have adverse consequences if issues of power, equity, and conflicting interests are not addressed.

Alongside these shifts in approaches to governance, food movements in Canada have also undergone a significant transformation in recent decades, moving from a primarily grassroots orientation to greater engagement in, and focus on, government relations and policy advocacy. Over the past three decades individuals, community groups, and organizations have responded to challenges in the dominant food system by establishing a range of programs and initiatives to both address the negative externalities of the dominant food system and continue the process of collaboratively rebuilding alternative systems based on values of health, equity, and sustainability (Wittman et al, 2011; Levkoe, 2014). Parallel to these efforts, food movements have advocated for government action and changes to key policies at the municipal, provincial/territorial, national, and international levels (Martin & Andrée, 2017; Koc et al, 2008; Wekerle, 2004). Food initiatives in Canada have been mobilizing through food movement networks since the late 1970s to collectively advocate for changes in the food system (Levkoe, 2015). Over time, as food movements evolved and matured, there has been an increasing focus on scaling-up local initiatives to address governance and food policy (see Figure 5.1).

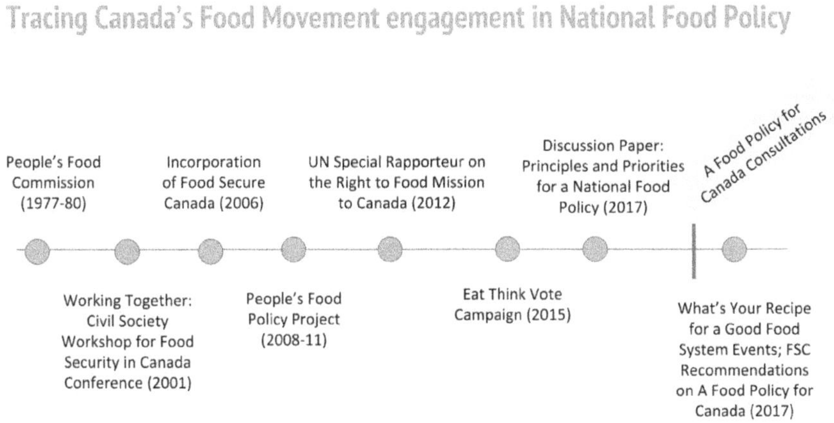

Figure 5.1 Food movements mobilization to impact food policy in Canada

One of the first collaborative efforts that brought together a range of civil society organizations using a food systems approach was the People's Food Commission. Established in 1977, the People's Food Commission was a broad grassroots initiative supported by a diversity of labour, anti-poverty, international development, and faith groups. It held hearings across the country with commissioners in each province to hear testimony from Canadians on the state of Canada's food system. At its core, the People's Food Commission sought to understand experience in the food system from consumers, producers, workers, and practitioners. *The Land of Milk and Money* was produced in 1980, summarizing the findings and identifying common challenges (PFC, 1980). Two decades later, Working Together: Civil Society Workshop for Food Security in Canada brought together 150 representatives from non-profit organizations, community groups, farmers, and government agencies to discuss avenues to increase Canada's commitment to food security both domestically and abroad (Koc & MacRae, 2001). From this gathering, came a mandate to establish a national food systems network. This paved the way for the establishment of FSC in 2006 as a "pan-Canadian alliance of organizations and individuals working together to advance food security and food sovereignty through three inter-locking goals: zero hunger, healthy and safe food, and sustainable food systems" (FSC, n.d.). In 2008, 30 years after the People's Food Commission, FSC played a major role in establishing the People's Food Policy project, culminating in an influential 2011 report that laid out key policy principles for a food system rooted in food sovereignty (PFP, 2011). Over the three years of the project, the People's Food Policy involved over 3,000 participants in community-based discussions, open forums, and collaborative working groups to discuss what was wrong with the food system and to envision policies that could lead to heath, equity, and sustainability (Levkoe & Sheedy, 2017).

Building from the People's Food Policy, food movement networks came together in advance of the United Nations Special Rapporteur on the Right to Food's 2012 visit to Canada. Supported by FSC and member organizations, the fact-finding tour included stops in major cities across Canada, meeting with civil society and government representatives to identify the state of Canada's food system and make recommendations to ensure all people have the right to food (UNGA, 2012). One of the report's key recommendations was the creation of a national food policy. Shortly thereafter, FSC's Eat Think Vote campaign, as part of the 2015 federal election adapted the People's Food Policy's participatory methodology to engage voters on five food policy issues (zero hunger, healthy school food, support for new farmers, and affordable food in the North). An overarching campaign priority was the development of a national food policy – to which all five major political parties in Canada signalled their support.

These activities demonstrate food movements' shift towards greater engagement with food policy and the state. The People's Food Commission (1980) explicitly situated itself separate from formal government process writing, "We called it the People's Food Commission because this was not going to be a government affair. We were going to hear directly from people and go back to them with our results" (p. 8). In contrast, the People's Food Policy and Eat Think Vote had more targeted goals of engaging with government policy as part of food sovereignty and a vision for a more sustainable and democratic food system. When the national food policy consultations were launched in 2017, FSC welcomed the announcement seeing a central role for civil society engagement. In a press release issued on May 29, 2017, the same day the consultations were launched, FSC wrote, "This is our chance to take concrete actions to reform our food system, from farm to fork. Our members across the country are excited to make their voices heard" (FSC, 2017d). This background helps to contextualize some of the opportunities food movement organizations navigated within the 2017 national food policy-building processes. Before discussing the food movement's engagement with the national food policy consultations, we turn briefly to a description of the concept of prefiguration in respect to the policy process.

Prefiguration and the policy processes

Prefiguration is an orientation to social change that focuses on enacting the desired future in the present (Siltanen et al, 2014). It is a strategic practice that seeks to close the gap between the real and the ideal, understanding "means and ends" as inherently linked and inseparable from one another (Maeckelbergh, 2011). As an ongoing process of imagining, doing, and reimagining alternative visions of society, prefiguration is, as van de Sande (2013) notes, inherently "experimental and experiential" (p. 232). From a prefigurative perspective, it is not just what we achieve that matters, it is also the process through which we achieve it.

Yates (2015) identifies two different ways in which prefiguration has been taken up in the literature, first as a general reference to the construction of alternatives (i.e. prefigurative outcomes), and second to the dynamic of how activities are performed (i.e. prefigurative processes). In this chapter, we take up the latter description through exploring food movements' engagement with the government's food policy consultation through prefiguration as a set of processes and practices to manifest the desired future in the present. For example, a prefigurative politics can be identified in the development of social movement networks through modeling the desired social relations. Juris (2008) writes about networks as "generating social laboratories for the production of alternative democratic values, discourses, and practices" (p. 3). Similarly, in an analysis of environmental justice movements, Schlosberg, (1999) writes, "networks are not simply a means to an end . . . they are an example of an attempt at an alternative political structure" (p. 142).

Dinerstein (2017) suggests that some civil society proposals are "untranslatable" into government policy, depending on the specific social, political, and economic conditions. If existing state institutions and socio-economic structures are based on an economic rationality directly hostile to the core values espoused by social movements (as is often the case in the neoliberal context), the implementation of policy proposals not in the "language" of government is an uphill battle. However, she suggests that this challenge of translation can be overcome through a prefigurative practice. Social movements can prefigure principles and values that are not yet "translatable" into policy content through the policy-building process. Dinerstein sees an emancipatory potential within policy-building processes, where social movements work to co-create and construct policy, thereby bringing alternative futures into being through the practice of policy-making. For example, for some, the language of food sovereignty is politically unrealistic as a policy proposal within the current Canadian government context. However, food sovereignty as a practice, can be modeled and experimented by food movement actors within the policy-making process.

However, Siltanen et al (2014) remind us of the need to be specific about what kind of alternative is being prefigured: "Prefiguration is not by definition positive, and its embedded vision of political processes and goals is open to evaluation and critique" (p. 276). Indeed, prefiguration in an abstract sense is quite vague; any number of possibilities could be prefigured. In the context of the food movement, prefiguration can be described as building a healthy, equitable, and sustainable food system through processes that are in and of themselves transformative. In a study of regional food networks in Canada, Levkoe (2015) provides insight into the "instrumental and prefigurative value of networks and the ways they can be a model of, and model for, establishing democratic governance structures, building new institutions, and engaging in different kinds of social relations" (p. 182). Prefiguration is a particularly useful concept to employ in an analysis of social movements engagements in food governance because it focuses attention on policy processes and challenges us to consider the space that can be created through these mechanisms not solely as a means to a predetermined end, but as a site of possibility and transformation beyond a specific policy outcome.

Developing a Food Policy for Canada

In this section, we describe the ways that food movement organizations, supported by FSC, participated in the development of a Food Policy for Canada between May and September in 2017. We focus on the specific activities that were undertaken during the consultation period to highlight how space was created by FSC to prefigure a systems-based policy approach and model participatory forms of governance.

"Have your say": The Canadian government's food policy consultations

Following the announcement of the national food policy initiative, the public consultation was set to last only four months. The official process consisted of an online survey (initially open for 60 days and extended for an additional month following a strong public response), an invitation only two-day Food Summit in Ottawa, a series of regional consultations held in five cities across the country, and the option to submit a formal Policy Brief. The Ottawa Food Summit brought together approximately 300 participants from diverse sectors across the country, including a strong representative group from civil society. This was a major event that enabled detailed discussions about the concerns and recommendations from the different stakeholders. The regional consultations were smaller events targeting specific sectors. The government initially committed to releasing a "What We Heard" report in the fall of 2017, summarizing the input received, with plans to have a draft policy by the summer of 2018. The release of the report was delayed several times; and at the time of writing there was no clear timeline for its release or the release of the draft policy.

As part of the national food policy consultation process the government predetermined four key themes: 1) increasing access to affordable food; 2) improving health and food safety; 3) conserving our soil, water, and air; and, 4) growing more high-quality food. The themes served as guidelines for the consultation tools, structuring both the online survey and the format of the Ottawa Food Summit.

Food movement participation in the food policy consultation process

In response, and in parallel, to the government's consultation processes, FSC, in collaboration with member organizations and allies, engaged in a range of policy-building activities. Working with individuals, academics, grassroots groups, and non-profit organizations, space was created to more meaningfully engage in the national food policy consultations. More than just filling in the survey or showing up at the consultation events, we argue that FSC took the opportunity to create a different kind of policy engagement that prefigured a set of processes and practices to model participatory and democratic policy engagement. FSC's orientation can be understood as a dual process of policy-building and movement-building: advancing strategic policy objectives in a manner that also strengthened the food movement and articulated a vision of food system transformation. This involved creating a series of opportunities for diverse actors to understand the government's approach and contribute their own experiences, ideas, and needs. The intention behind these initiatives was to make the process more inclusive and accessible to social movement organizations across Canada, and to build stronger relationships and connections within the food movement.

Table 5.1 outlines the FSC-led initiatives organized into three categories along with their associated activities and outcomes: 1) strategic policy principles to seed food system transformation; 2) capacity-building activities to enhance food movements' ability to engage in policy-building processes; and, 3) collaborative policy-making infrastructure that sought to model multi-stakeholdism and co-governance in order to build consensus on key ideas and proposals. These activities are discussed in the following subsections.

Table 5.1 FSC-led initiatives for the national food policy

Category	*Activity*	*Outcomes*
Strategic policy principles	*From Patchwork to Policy Coherence: Principles and Priorities of Canada's National Food Policy*	a discussion paper synthesizing literature on key food system issues and identifying key principles and priorities to frame the discussion on national food policy; released in advance of the government's formal consultation process
	Policy Brief on a Food Policy for Canada	summary of food movements' priorities; FSC's formal submission to the government's consultation's process
	Mapping the food policy landscape in Canada	a policy scan, presented through six discussion papers; peer reviewed by practitioners and academics
	Food policy briefing notes	a series of briefing notes submitted by FSC members and allies addressing key food policy issues
	Canadian Food Studies Journal Special Issue on Building an Integrated National Food Policy for Canada	co-edited by FSC, the special journal issue included original research papers and commentaries
Capacity- and movement-building	Workshops at the Annual Meeting of the Canadian Association for Food Studies	a series of policy-related workshops including a meeting of over 30 academic and community leaders to share reflections on the engagement process and discussion of next steps (May 2017)
	Civil Society Pre-Food Summit Meeting	a gathering of more than 60 food movement representatives to strategize in advance of the Ottawa Food Summit (June 2017), including the creation of an online network to share background/briefing materials and facilitate conversation
	Webinars/conference calls	regular webinars and conference calls with organizational and individual members and allies to provide key updates on engagement activities and solicit feedback
	Five Big Ideas for a Better Food System	a set of policy proposals framed around five themes

Collaborative infrastructure	Recipe for Good Food System events	hosted 23 in-person and four online discussions (organized by theme); produced an event guide along with a final report synthesizing the common issues, themes, and priorities
	Ad Hoc Working Group on Food Policy Governance	a multi-stakeholder working group of civil society, academics, and industry groups that outlined a vision for a National Food Policy Council
	Working Group on the Right to Food	a working group of human rights NGOs and lawyers/legal experts that outlined a shared vision strategy to implement the right to food in Canada

Prefiguring a systems-based, participatory food policy approach

In contrast to the government's consultation process, FSC's initiatives were not solely focused on deriving particular policy outcomes. Beyond collecting valuable information to include in contributions to the national food policy, the initiatives attempted to be prefigurative in nature. The activities outlined in Table 5.1 were not just about lobbying to win specific policy objectives, they were also intended to grow the food movement, and use the space created by the consultation process to strengthen relationships, build capacity, and experiment with alternative models of governance. In other words, FSC and its allies actively experimented with different participatory processes to enact future food system possibilities. These experiments were rooted in a practice of food sovereignty to prefigure an integrated system-based approach to food policy based on participatory, democratic, and collaborative decision-making.

While on the surface the government-outlined themes appeared to move food systems discussions forward, many civil society actors questioned the framing of these priorities and the logic of breaking down food policy into single issue categories. FSC allies also criticized the language used within the details of the four themes. For example, FSC's report-back on the Ottawa Food Summit noted that "addressing food insecurity is not simply a question of affordability or market access, but rather an issue of human rights, Indigenous food sovereignty and vitally important income supports" (FSC, 2017c). Shortly after the launch of the national food policy consultations, FSC released Five Big Ideas for a Better Food System which included: Realizing the Right to Food; Championing Healthy and Sustainable Diets; Supporting

Sustainable Food Systems; Making Food a Part of Reconciliation; and, Inviting More Voices to the Table. Drawing on previous policy goals from the 2015 Eat Think Vote campaign, the People's Food Policy, and reflecting the core mission and values of FSC, these five themes identified a series of collective priority outcomes from the food movement. While the four themes put forth by the federal government addressed food system issues individually, FSC's Five Big Ideas presented interrelated priorities rooted in a vision of healthy, equitable, and sustainable food systems. For example, Championing Healthy and Sustainable Diets prioritized food that is healthy for both humans and the natural environment, while Invite More Voices to the Table challenged the government to adopt a more inclusive approach to both the consultation process and food policy governance. Perhaps most importantly, FSC's Five Big Ideas explicitly referenced the need for reconciliation in discussions on national food policy; a glaring absence in the government's official documents for a Food Policy for Canada.

A systems-based approach was also prefigured through the creation of spaces for discussion and collaboration that intentionally brought together a range of different perspectives within food movements. For example, the caucus FSC hosted prior to the Ottawa Food Summit organized participants into breakout groups that saw dieticians, Indigenous people, food bank workers, and academics collectively identify common goals and strategies. Further, the discussions that produced FSC's food policy briefing notes and other contributions to the broader debate actively engaged different ideas, including the tensions that emerged. This cultivated a food systems approach by widening the perspectives of individuals to include the experiences of a wide range of actors. Moreover, the processes that FSC used to engage its members and allies in critical discussions represented a different approach to how civil society is typically engaged by governments in policy development. Instead of simply responding to a set of predetermined themes (as the government's consultation did), FSC attempted to model a different kind of policy-making process that engaged a broader diversity of communities and encompassed a wider set of priorities rooted in issues of health, equity, and sustainability over productivity and economic profit.

Throughout the consultation period, FSC developed several participatory decision-making processes that prefigured a more inclusive approach to policy-building. In addition to naming participatory and inclusive governance as one of its priorities for the national food policy, FSC attempted to model participatory decision-making processes in its outreach and engagement activities. This included soliciting feedback from its members on the *From Patchwork to Policy Coherence* discussion paper and the final Policy Brief on national food policy. For example, the discussion paper was released in anticipation of the government's consultation process, as a primer for its members and allies. It synthesized previous research, reports, and experiences from Canadian food movements to provide context, frame the conversation, and identify key challenges and questions. The paper went through a feedback process with FSC members, with almost 30 different reviews from both academics and community-based practitioners. Through this review process, FSC was able to identify points of consensus and tension among the diverse perspectives that make up food movements and articulate a shared vision for Canada's food system. It also created a feedback loop between FSC members, representing a range of food movement actors and networks, and FSC staff, enabling a more inclusive and representative set of policy priorities.

There was also an attempt to prefigure multi-stakeholder governance between a broader set of non-state actors. The Ad Hoc Working Group for Food Policy Governance was an eclectic mix of civil society organizations who sought to work together to achieve consensus on the question of governance within the Food Policy for Canada. The five core members, FSC, the Canadian Federation of Agriculture, Maple Leaf Foods, the Arrel Food Institute, and the McConnell Foundation, were joined by a long list of food movement organizations and more mainstream food and agriculture groups in calling for the creation of a National Food Policy Council (NFPC) as part of the Food Policy for Canada to, in their words, "proactively engage with these diverse stakeholders to provide ongoing input into the implementation of A Food Policy for Canada" (Ad Hoc Working Group for Food Policy Governance, 2018). The process to create the proposal for a NFPC was in and of itself prefiguring the very outcome it wanted to achieve, collapsing the distinction between means and ends.

These attempts at inclusivity were also evident in the mobilization activities leading up to the Ottawa Food Summit. While FSC is a member-based organization, the briefing materials and caucus meeting were open to any civil society or academic representative, regardless of their membership status. This enabled a greater diversity of food movement actors to share their perspectives on policies they desired and engage in critical discussion about the key themes put forward by the government. In particular, this was an important opportunity to take direction from Indigenous communities and representatives from food movement organizations that expressed feelings of marginalization from the government-led processes.

While the impetus for many of these activities was to facilitate and support civil society contributions to the government's consultation process, prefiguring a systems-based, participatory food policy approach was about much more than strictly engaging with the state. In practice, the activities that fed into these processes had a variety of purposes to achieve multiple goals, of which policy-making was only one. In other words, FSC members and allies approached the policy-building process as both a means and an end. While there was a concerted effort to influence federal food policy, the activities and outputs also had the objective of strengthening food movements through network building, enhancing capacity, and critical engagement. Particularly for those critical of the policy-development process, and those skeptical of its outcomes, this "inside–outside" approach to movement-building and policy impact became an important rationale for engagement.

Beyond just focusing on contributions to the Food Policy for Canada, a key aspect of FSCs engagement was to strengthen Canada's food movements. This was evident in FSC's attempts to revive existing partnerships and facilitate new connections among its member organizations and other allies. While each of the participating food initiatives was rooted in place and held significant experiential knowledge, it was the connections made between organizations and sharing of common (and differing) ideas that created additional capacity for action. Through briefing papers, webinars, workshops, and support for community engagement events, FSC attempted to create spaces for knowledge-sharing and skill-building among the participants. This also served as an opportunity for strategic mobilization of movement activities and experiences for impacting policy at multiple levels. For example, through the Recipe for a Good Food System events, FSC partnered with local and provincial/territorial community groups to hold 23 community engagement events. The events created open space for discussion and collaboration at the community level, all of which were synthesized into a formal submission to the food policy consultation process. However, the events also went far beyond providing input into a government consultation process. In FSC's report synthesizing the feedback received through the Recipe for a Good Food System events, they noted that "in some communities the events were a 'spark' for continuing local dialogue and action around food" (FSC, 2018). Here the dual purpose of strategic policy intervention and movement-building is clearly evident: FSC used the policy opening created by the government consultation period to generate policy proposals from grassroots food system actors, but also to strengthen relationships with partners and allies and forge new connections.

Some of the activities outlined above also created space to tackle difficult conversations and areas of tension. FSC member conference calls, as well as the Pre-Ottawa Food Summit Workshop were opportunities to voice concerns and discuss strategy and tactics. For example, food security advocates and ecological farmers were able to discuss the need for fair wages and economic justice across the food system, in contrast to the call for more affordable food put forth by the government's policy themes. By coordinating these kinds of engagement activities and difficult discussions, FSC contributed to creating space and capacity for critical analysis and developing framings that addressed common concerns and underlying problems within the dominant food system. Further, these engagements served as an experimentation with models for collaboration beyond the limited opportunities for formal contribution.

Limits to prefiguring food sovereignty policy

In the above section, we argue that FSC made a concerted effort to prefigure participatory policy-building processes and made some significant advances. However, these intentions to facilitate food movements' various concerns and contributions came up against a series of challenges including major pressures to conform to the government's predetermined themes, language, and processes, as well as resource constraints. At times, this meant that the goals of food sovereignty and deeper food systems transformation were sidelined in an attempt to present the movement as a legitimate voice and contributor to the process. This resulted in a key tension that many civil society participants recognized and openly discussed as part of the various engagements. Some were pulled towards maintaining more radical positions that were clearly outside the government's purview, while others attempted to work within the predetermined process.

The government's stated desire to "hear from Canadians" stands in contrast to many years of successive governments that had little interest in listening to the needs of food movement organizations, beyond corporate interests. While this was undoubtedly harmful to Canada's food system, there is a certain liberty in having a government that is openly hostile to grassroots objectives, as there is no pressure to pander to the state. In contrast, once a formal invitation is made by government, there is a sense of obligation to take advantage of the opening, regardless of how narrow it may be. However, engaging in these government-led and defined processes can have the effect of diluting more radical politics. For example, a comparison of FSC's 2017 discussion paper on national food policy and the 2011 People's Food Policy illustrates a strategic shift in language in some key areas. Most notably, there was a decreased emphasis on food sovereignty and an increased discussion of multi-stakeholder governance mechanisms as opposed to civil society taking power and playing a central role in decision-making.

Narrowing the horizons of possibilities

To be able to participate in the government's process, food movement organizations were forced to make a series of trade-offs between the three factors relating to the execution of power (see Introduction in this volume): access to instrumental resources vs. marginalization, a transformative vs. reformist orientation, and wrestling control from others vs. working collaboratively. While there is much to gain, engaging in these processes also risks alienating groups and issues that don't fit the government's predetermined process (e.g. Indigenous food sovereignty, migrant farm worker rights). Further, participating in the government's process risks legitimizing a process that was flawed from the outset. In addition, participation in these policy-building processes may push food movement actors to engage in activities or collaborations for strategic purposes that they would otherwise not consider. For example, instead of approaching policy from a food sovereignty perspective, they are forced to reframe priorities to match the structure of the state. While a practice of food sovereignty, as discussed earlier, may partially overcome this limitation, there is nonetheless a power, and a significance, in explicitly naming and articulating a food sovereignty politics.

Siltanen et al (2014) cautioned that prefiguration can be used for a variety of purposes, to prefigure a diversity of future possibilities, not all of which may be seen as desirable. For example, the multi-stakeholder process behind the Ad Hoc Working Group on Food Policy Governance was seen as a strategically important opening to present a unified front on crucial issues, namely the question of governance within a national food policy. Agreement on a shared set of recommendations[6] was seen as a major achievement by some, while others expressed concern that such an approach represented a shift away from FSC's historical roots and grassroots orientation.

While recognizing the strategic importance of providing input in the form of survey responses, attending the Ottawa Summit and contributing recommendations, participation at any level signals tacit approval and consent to the process – and in turn the final outcome. Given the accelerated timeline and the pre-existing themes, there are questions as to how and to what degree the input provided through these consultation mechanisms will actually shape the development of the government's food policy. FSC was given a responsibility to be a voice of food movements but at the same time, the organization was under-resourced and had little capacity to play a substantial role. With such a lack of resources, FSC was limited in the degree to which they could be part of the process. Initial conversations included ambitious plans for a multi-year engagement process, requiring over $1 million in funding to facilitate policy labs, collaborative workings groups, and cross-country community consultations. In the end, a much smaller poll of resources was available, mostly small sums contributed by food movement organizations, a small contract from Agriculture and Agri-foods Canada to support the Recipe for a Good Food System events, activities within existing academic research partnerships, and a Mitacs postdoctoral fellow position. Despite the time and resource constraints, the list of activities laid out in Table 5.1 represents a significant undertaking, and the fact that they still took place is a testament to the commitment of many food movement activists to making the most of this policy opening.

When FSC brought together a group of more than 30 representatives from food movements across Canada in May 2017 at the annual meeting of the Canadian Association for Food Studies, there was an overwhelming sentiment of disappointment that the government's proposed consultation process fell substantially short of expectations and desires. Participants believed this was a significant opportunity to influence policy but that it would be extremely difficult with only four months. There was also critique about the limited nature of the kinds of feedback accepted. Based on conversations with member organizations, FSC observed:

> The exercise [of filling out the survey] can be frustrating for people who believe in a joined-up food policy – the very essence of which is to tackle health, environment and equity considerations together, not to trade one off against the other. It's not a question of choosing

> between a priority for "Making nutritious food more available to all Canadians" or "Supporting the growth of local and regional food production," instead it's important how they can both be addressed, potentially together.
>
> (FSC, 2017b)

In the context of increasing calls to "Have your say" and contribute to government consultation processes, the invitation to participate does not necessarily lead to meaningful participation. We must acknowledge barriers to participation such as limited time, access to resources and knowledge, as well as the cultural and social capital bound up in these consultation processes. In the case of the Ottawa Food Summit, travel and accommodation costs of participants were covered, but those attending still required time off work (if they were not attending on behalf of their employer) and the mobility to disengage from any domestic labour with relatively short notice. While many food movement organizations were able to participate, there were also many people that could not provide feedback into the consultation process.

This raises an important question of how power was mobilized within the consultation process. There were several symbolically important moments of "speaking truth to power"; for instance, when several Indigenous people got up to the mic at a plenary session during the Ottawa Food Summit and reminded the government representatives of their treaty obligations and commitment to nation-to-nation relationships. However, these displays of symbolic power were largely overshadowed by the institutional and structural power held by the state, and to a lesser degree, prominent social movement organizations (including FSC) to shape the overall discourse and horizon of possibilities. In reflecting on the discourses that shaped the consultation process, a distinction needs to be made between including the concerns and perspectives of marginalized communities (Indigenous and racialized people, migrant workers, those with lived experience of food insecurity) and centering the voices of these communities and taking leadership and direction from their own goals and objectives. This may mean the difference between a Food Policy for Canada that includes an acknowledgment of the importance of reconciliation and a decolonizing food policy; a Food Policy for Canada that seeks to tackle food insecurity and a food policy grounded in food justice. Particularly for social movement organizations, it is often in the negotiation of power where the tensions between strategic policy reforms and deeper food systems transformation are laid bare.

These kinds of tensions have been identified elsewhere in respect to food sovereignty movements engaging with state policy. Other such engagements have raised questions whether food sovereignty is "translatable" into a federal government food policy. Wittman (2015) identifies several challenges for institutionalizing food sovereignty in her discussion of Brazilian food policy. The first is the question of scalability from the ground up: how to implement food sovereignty at a national level while still maintaining the principles of democracy and place-based solutions. The second challenge is impositions from the top-down: how to engage with an international trade regime that is fundamentally hostile to principles of food sovereignty. If food sovereignty is "ultimately about changing and decentralizing power in the food system" (Wittman, 2015, p. 179), we question whether it is even possible for a national food policy, developed by government, to further its implementation. Wittman (2015) however, suggests that institutionalizing food sovereignty at different scales demands redistributing resources and power within the food system. She points to the creation of an enabling environment for "autonomous and localized" initiatives through legislative reforms, public procurement policies, and global-level dialogue on trade and food security.

Recognizing these limitations, prefiguration provides an important avenue forward, whereby food sovereignty can become institutionalized through practice – through policy *processes* rather than solely policy *outcomes*. Place-based initiatives and food movement organizations can advocate for food sovereignty through the deliberate practice and modeling of food sovereignty principles in the way they engage in policy-building processes. In this way prefiguration can be a tool for managing the tension and complexity of working both "inside" and "outside" the state (see, for example Siltanen et al, 2014, p. 261). Prefiguration in policy development processes can act as an insulating force to pressures of cooptation and conformity; creating a space where social movement organizations and actors can engage with the state on their own terms.

Policy engagement as prefiguration might be seen as a kind of power deployed by food movement actors. The power cube described in Chapter 1 brings light to these processes in respect to the ways that spaces of engagement are created. Gaventa (2006) reminds us that governance spaces are not neutral and as a result, social movements face a series of closed and invited spaces. The lens of prefiguration helps to highlight the dynamic nature of power within these policy spaces. Rather than existing as static spaces that are closed, invited, or claimed, these labels are contested and negotiated. In the case of the government-led consultations on a Food Policy for Canada, food movement organizations were invited into the space, but they were also able to, temporarily, claim space, by using it for a purpose other than that for which it was originally intended, and by pushing the boundaries of what was under discussion. However, at the same time the invitation into this space was somewhat deceptive in respect to how effectively it could actually be used to further the goals of food movements. As a result, there are new opportunities, tensions, and internal challenges that food movement actors will be forced to navigate moving forward.

A prefigurative practice can be used as a means of claiming space, and can help to insulate social movements from these challenges, but it is unlikely that they will be avoided altogether. In practice, prefigurative processes are complex and forced to interact, to some degree, with outside forces. Social movement actors, including FSC and its allies, are undoubtedly shaped and influenced by the structural forces and large-scale trends, yet at the same time, the activities and orientations of food movement organizations engaged in national food policy go beyond an attempt to cope with and respond to current realities.

Conclusions: Balancing process and outcome in policy development

While the development of a national food policy is a significant step forward in addressing food system challenges, it is also important to critically examine the process of its development, in particular the quality and heft of the consultation and the degree to which meaningful spaces of participation are created. While there is reason for caution, we suggest that a prefigurative practice can contribute to a more meaningful and purposeful engagement in policy-building processes, where both the means and ends of policy-building are considered. Ultimately, an effective approach to policy as prefiguration requires the advancement of both "policy as outcome" and "policy as process". If the government's Food Policy for Canada that develops out of this process does not provide any substantive progress on the pressing issues facing our food system, it will be hard to justify the amount of time and resources that has been dedicated to the consultation process, regardless of the degree to which food movements were strengthened as a result. At the same time, if the Food Policy for Canada is seen as a victory but came at the expense of a narrowing of possibilities and reduced grassroots engagement, we would argue it is not a victory worth celebrating.

Civil society is increasingly being asked to articulate its concerns, but also to outline concrete solutions and fully developed policy proposals complete with budget allocations and implementation strategies. In the case of the national food policy consultations, this provides an opportunity to develop governance relationships where civil society had an opportunity to provide input into government-led policy processes. However, as long as the state ultimately controls the process and without meaningful opportunities to provide input along with democratic and participatory participation, the processes feel more like tokenism than engagement (also see Arnstein, 1969). Thus, we suggest that the consultation can be understood as a point of tension with a desire for a more inclusive policy engagement process by food movements paralleled by a pull towards multi-stakeholderism. As the Food Policy for Canada has yet to be released, it is difficult to fully assess the degree to which FSC's approach was successful in making concrete policy gains while also prefiguring deeper food system transformations. It remains to be seen whether the government's desire to consult with Canadians will lead to further opportunities of collaboration and co-governance. However, regardless of the policy outcome, through the policy-building process, FSC along with its member and allies, was able to create a space of collaboration and imagination that was at least partially successful in prefiguring a food future worth fighting for.

Acknowledgements

The authors would like to thank the staff and members of Food Secure Canada, as well as other food movement organizations and activists, for their openness to collaboration and their commitment to imagining and building a better food system. The authors are also grateful to fellow FLEdGE (Food: Locally Embedded Globally Engaged) participants for their helpful feedback and comments on previous versions of this chapter, to CFICE (Community First: Impacts of Community Engagement) for creating a space in which to explore and experiment with innovative forms of community–academic collaboration, and finally to SSHRC (Social Sciences and Humanities Research Council) for providing financial support to both the aforementioned research partnerships.

Notes

1 We describe social movement organizations as those identifying with a social movement or countermovement and actively involved in mobilizing individuals and other organizations towards changing or restructuring society (Lofland, 1996). These organizations are often a subset of civil society organizations, which encompass a wider array of organizations.
2 While many food movements' activities that engage with policy have their roots in much broader and historical processes, in this chapter, we focus our attention on the May–September 2017 time period to capture the specific set of actions and mobilizations that took place in relation to the government's consultation process.
3 For more on the Canadian government's national food policy process, see www.canada.ca/en/campaign/food-policy.html
4 CFICE is a partnership-based action research project that studies how academics and practitioners work together to positively impact their communities. For more information see www.carleton.ca/communityfirst/
5 See http://pm.gc.ca/eng/minister-agriculture-and-agri-food-mandate-letter
6 An open letter outlining the recommendations to Minister of Agriculture and Agri-food Lawrence MacAulay was signed by over 40 food and agriculture organizations. Copies of the letter and report are available at: https://foodsecurecanada.org/policy-briefs-and-reports

References

Ad Hoc Working Group for Food Policy Governance (2018) *The case for a national food policy council*. Available at: https://foodsecurecanada.org/sites/foodsecurecanada.org/files/attached_files/the_case_for_a_national_food_policy_council_final.pdf

Alkon, A. H. & Guthman, J. (eds) (2017) *The new food activism: opposition, cooperation and collective action*. Oakland, University of California Press.

Andrée, P., Cobb, M., Moussa, L., & Norgang, E. (2011) Building unlikely alliances around food sovereignty in Canada. *Studies in Political Economy*, 88(1), 133–159.

Arnstein, S. R. (1969) A ladder of citizen participation. *Journal of the American Institute of Planners*, 35(4), 216–224.

Barling, D., Lang, T., & Caraher, M. (2002) Joined–up food policy? The trials of governance, public policy and the food system. *Social Policy & Administration*, 36(6), 556–574.

Beaulieu, M. (2015) *Demographic changes in Canadian agriculture*. Statistics Canada. Available at: www.statcan.gc.ca/pub/96-325-x/2014001/article/11905-eng.htm

Candel, J. J. & Pereira, L. (2017) Towards integrated food policy: main challenges and steps ahead. *Environment Science and Policy*, 73, 89–92.

Constance, D. H., Renard, M. C., & Rivera-Ferre, M. G. (2014) *Alternative agri-food movements: patterns of convergence and divergence*. Bingley, Emerald Group Publishing.

Dinerstein, A. (2017) Co-construction or prefiguration? The problem of the "translation" of social and solidarity economy practices into policy. In: North, P. & Scott Cato, M. (eds) *Towards just and sustainable economies: The social and solidarity economy north and south*. Bristol, Policy Press, pp. 57–71.

Environment Canada (2014) *Canada's emission trends 2014*. Government of Canada. Available at: www.canada.ca/en/environment-climate-change/services/climate-change/publications/emission-trends-2014.html

FSC (Food Secure Canada) (n.d.) *What we do*. Available at: https://foodsecurecanada.org/who-we-are/what-we-do

FSC (2017a) *A primer on national food policy in Canada*. Available at: https://foodsecurecanada.org/resources-news/news-media/national-food-policy-primer

FSC (2017b) *Make your voices heard on national food policy*. Available at: from https://foodsecurecanada.org/resources-news/news-media/make-your-voice-heard-national-food-policy

FSC (2017c) *Food movement gathers momentum at Ottawa Food Policy Summit*. Available at: https://foodsecurecanada.org/resources-news/news-media/food-movement-gathers-momentum-ottawa-food-policy-summit

FSC (2017d) *FSC's Press release on national food policy consultations*. Available at: https://foodsecurecanada.org/resources-news/news-media/press-releases/food-movement-welcomes-consultations-national-food-policy

FSC (2018) *Final report: civil society engagement towards a food policy for Canada*. Available at: https://foodsecurecanada.org/resources-news/resources-research/report-civil-society-engagement-towards-food-policy-canada

Gaventa, J. (2006) Finding the spaces for change: a power analysis. *IDS bulletin*, 37(6), 23–33.

Gooch, M. & Felfel, A. (2014) *$27 billion revisited: The cost of Canada's annual food waste*. Value Chain Management International Inc. Available at: http://vcm-international.com/wp-content/uploads/2014/12/Food-Waste-in-Canada-27-Billion-Revisited-Dec-10-2014.pdf

Holt-Gimenez, E. (ed.) (2011) *Food movements unite! Strategies to transform our food system*. Oakland, Food First Books.

Ilcan, S. & Basok, T. (2004) Community government: voluntary agencies, social justice, and the responsibilization of citizens. *Citizenship Studies*, 8(2), 129–144.

Institute for Health Metrics and Evaluation (2010) *Global burden of disease study profile: Canada*. Available at: www.healthdata.org/sites/default/files/files/country_profiles/GBD/ihme_gbd_country_report_canada.pdf

IPES-Food (International Panel of Experts on Sustainable Food Systems) (2016) *From uniformity to diversity: a paradigm shift from industrial agriculture to diversified agroecological systems*. International Panel of Experts on Sustainable Food systems. Available at: www.ipes-food.org

Isin, E. (2000) Governing cities without citizens. In: Isin, E. (ed.) *Democracy, citizenship and the global city*. London, Routledge, pp. 148–168.

Jessop, B. (2002) Liberalism, neoliberalism, and urban governance: a state-theoretical perspective. *Antipode*, 34(3), 451–472.

Juris, J. S. (2008) *Networking futures: the movements against corporate globalization*. Durham, Duke University Press.

Koc, M. & MacRae, R. (eds) (2001) *Working together: civil society working for food security in Canada*. Toronto, Media Studies Working Group.

Koc, M., MacRae, R., Desjardins, E., & Roberts, W. (2008) Getting civil about food: the interactions between civil society and the state to advance sustainable food systems in Canada. *Journal of Hunger & Environmental Nutrition*, 3(2/3), 122–144.

Lang, T., Barling, D., & Caraher, M. (2009) *Food policy: integrating health, environment and society*. Oxford, Oxford University Press.

Levkoe, C. Z. (2011) Towards a transformative food politics. *Local Environment*, 16(7), 687–705.

Levkoe, C. Z. (2014) The food movement in Canada: a social movement network perspective. *Journal of Peasant Studies*, 41(3), 385–403.

Levkoe, C. Z. (2015) Strategies for forging and sustaining social movement networks: a case study of provincial food networking organizations in Canada. *Geoforum*, 58, 174–183.

Levkoe, C. Z. & Sheedy, A. (2017) A people-centred approach to food policy making: lessons from Canada's People's Food Policy project. *Journal of Hunger & Environmental Nutrition*. Available at: www.tandfonline.com/doi/abs/10.1080/19320248.2017.1407724?journalCode=when20

Levkoe, C. Z. & Wakefield, S. (2014) Understanding contemporary networks of environmental and social change: complex assemblages within Canada's "food movement". *Environmental Politics*, 23(2), 302–320.

Lofland, J. (1996) *Social movement organizations*. New York, Aldine de Gruyter.

MacRae, R. (2011) A joined-up food policy for Canada. *Journal of Hunger & Environmental Nutrition*, 6(4), 424–457.

MacRae, R. & Winfield, M. (2016) A little regulatory pluralism with your counter-hegemonic advocacy? Blending analytical frames to construct joined-up food policy in Canada. *Canadian Food Studies*, 3(2), 140–194.

Maeckelbergh, M. (2011) Doing is believing: prefiguration as strategic practice in the alterglobalization movement. *Social Movement Studies*, 10(1), 1–20.

Martin, S. J. & Andrée, P. (2017) Putting food sovereignty to work: civil society governmentalities and Canada's people's food policy project (2008–2011). *Journal of Civil Society*, 13(4), 374–391.

McKeon, N. (2017) Are equity and sustainability a likely outcome when foxes and chickens share the same coop? Critiquing the concept of multistakeholder governance of food security. *Globalizations*, 14(3), 379–398.

Mitchell, K. (2001) Transnationalism, neo-liberalism, and the rise of the shadow state? *Economy and Society*, 30(2), 165–189.

Nyéléni (2007) *Declaration of the forum for food sovereignty*. Available at: www.nyeleni.org/spip.php?article290

Patel, R. C. (2009) Food sovereignty. *Journal of Peasant Studies*, 36(3), 663–706.

PFC (People's Food Commission) (1980) *The land of milk and money: the national report of the People's Food Commission*. Kitchener, Between the Lines.

PFP (People's Food Policy) (2011) *Food democracy and governance*. Available at: https://foodsecurecanada.org/sites/foodsecurecanada.org/files/DP10_Food_Democracy_and_Governance_0.pdf

Power, E. M. (2008) Conceptualizing food security for Aboriginal people in Canada. *Canadian Journal of Public Health/Revue Canadienne de Sante'e Publique*, 99(2), 95–97.

Rittel, H. W. J. & Webber, M. M. (1973) Dilemmas in a general theory of planning. *Policy Sciences*, 4(2), 155–169.

Schlosberg, D. (1999) Networks and mobile arrangements: organizational innovation in the US environmental movement. *Environmental Politics*, 8(1), 122–148.

Siltanen, J., Klodawsky, F., & Andrew, C. (2014) "This is how I want to live my life": an experiment in prefigurative feminist organizing for a more equitable and inclusive city. *Antipode*, 47(1), 260–279.

Statistics Canada (2014) *Canadian community health survey*. Available at: www.statcan.gc.ca/daily-quotidien/150617/dq150617b-eng.htm

Statistics Canada (2017) *2016 Census of agriculture*. Farm and farm operator data. Available at: www.statcan.gc.ca/pub/95-640-x/95-640-x2016001-eng.htm

UNGA (United Nations General Assembly) (2012) *Report of the Special Rapporteur on the Right to Food*. Olivier De Schutter, Mission to Canada. UN Doc A/HRC/22/50/Add.1.

van de Sande, M. (2013) The prefigurative politics of Tahrir Square: an alternative perspective on the 2011 revolutions. *Res Publica*, 19(3), 223–239.

Wekerle, G. R. (2004) Food justice movements: policy, planning, and networks. *Journal of Planning Education and Research*, 23(4), 378–386.

Wittman, H. (2015) From protest to policy: the challenges of institutionalizing food sovereignty. *Canadian Food Studies/La Revue canadienne des études sur l'alimentation*, 2(2), 174–182.

Wittman, H., Desmarais, A., & Wiebe, N. (eds) (2010). *Food sovereignty: reconnecting food, nature and community*. Halifax, Fernwood.

Wittman, H., Desmarais, A., & Wiebe, N. (eds) (2011) *Food sovereignty in Canada: creating just and sustainable food systems*. Halifax, Fernwood.

Yates, L. (2015). Rethinking prefiguration: alternatives, micropolitics and goals in social movements. *Social Movement Studies*, 14(1), 1–21.

6 Comparing the effectiveness of structures for addressing hunger and food insecurity

Molly D. Anderson

Introduction

Many countries have shown significant improvements in food security over the last few decades, yet in some countries progress has reversed. In July of 2017, the Food Security Information Network reported that hunger and food insecurity are increasing again (FSIN, 2017) in lockstep with national and regional conflicts and climate change; nearly 60 per cent of people suffering from chronic hunger live in countries struck by these two scourges. But for countries that are not feeling severe impacts of climate change and conflict, why have some shown success in reducing food insecurity and others have not? As explained in the introduction to this volume, attention has turned to food policy governance, or how decisions are made affecting hunger and food insecurity, to try to answer that question.

This case study compares governance structures for addressing hunger and food insecurity at different scales. My interest is in what happens when citizens have genuine ongoing opportunities for contributing to public policy about food systems, specifically about eliminating hunger and food insecurity. For this chapter, I examine the roles that civil society plays in organizations trying to reduce hunger and food insecurity, and circumstances in which they can increase effectiveness of these efforts. My primary hypothesis is that civil society participation radically changes the outcomes of deliberation in the direction of greater social and environmental sustainability than discussions would have taken if they had been dominated by business or government interests. A secondary hypothesis is that civil society participation, under the right circumstances, can open the door to food system transformation which facilitates access to healthy food for all as a human right.

This chapter is based on participant observation in two forums over the past five years – the Committee on World Food Security (CFS) and the Vermont Farm to Plate Network (F2P). I attended the week-long CFS meetings at the Food and Agriculture Organization (FAO) in Rome from 2013 through 2017 and the two-day Civil Society Mechanism (CSM) preparatory sessions before each meeting, in which up to 250 representatives from civil society organizations across the globe, plus a sprinkling of academicians, meet to strategize. I have also participated in several CSM Working Groups and an invited Technical Forum on Monitoring. In addition, I participate regularly in approximately monthly CSM-North America conference calls with civil society representatives and academicians from the United States (US) and Canada. This level of involvement over the last five years has allowed me to meet and work with the civil society representatives from many countries, as well as US delegates, some of the permanent FAO representatives from country governments, and FAO staff. I also participate regularly in the Vermont F2P Initiative, a ten-year state-wide food system strategy. I have attended two of the annual Summits that bring together diverse people across the state who are interested in food and agricultural issues, and helped to organize and lead Summit workshops. I have been the co-chair of the Research Crosscutting Team for the past two years, attend meetings of the Food Access Crosscutting Team, and participate in calls.

While focusing on the CFS and F2P, I briefly consider two other decision-making structures that deal with hunger or food access and which have been recognized as innovative in academic literature and the popular press: the Brazilian National Council on Food and Nutrition Security (CONSEA) and food policy councils in the US and Canada. I selected these examples to fill in additional municipal/regional and subnational scales of decision-making. Each model in the case study prioritizes policy on hunger and food insecurity, so they allow comparisons of effectiveness in meeting that goal. Moreover, each model has been in place long enough (at least eight years) to be able to demonstrate successes, and was created as a lasting structure. They were all created in circumstances where the need for participation from multiple perspectives and backgrounds, including representatives of civil society, was recognized.

Below, I describe each structure briefly, explaining how and by whom it was created and for what purpose(s). Then I describe some of the documented achievements that each organization has accomplished, and challenges they face. The discussion section sets forth hypotheses of attributes that contribute to high functionality of civil society participation in multi-actor organizations.

A foundational question is: Who is civil society and does it truly advocate for the public good? In the realm of food systems, Marsden et al (2010) divide civil society into those holding social interests (e.g. child labour, fair trade, occupational health, public health); environmental interests (e.g. biodiversity, sustainability, insecticides); and consumer interests (e.g. food safety, food security, choice, price). Each of these interest areas can be represented by social movements and non-governmental organizations (NGOs), and they can fall across the political spectrum from ultra-conservative to radically progressive. That is, civil society includes neo-Nazi groups and Tea Party activists who try to suppress progressive changes in society. For the purposes of this case study and in alignment with other chapters in this book, however, civil society is considered to consist of social movements and NGOs that are working for greater justice and sustainability, with particular attention to hunger and food insecurity.

Civil society participation in eliminating hunger and food insecurity, and related food system issues, is important yet contentious. Advantages of civil society participation include the following. First, it gives greater legitimacy to decisions, since it is clear that multiple perspectives including the public interest have been considered, which enhances the public trust often lacking in work on food issues (Ankeny, 2016; Scholte, 2001). Second, civil society participation enables those who are suffering from hunger and food insecurity (i.e., violations of their right to food and adequate nutrition) to be heard from directly and to contribute to creating solutions appropriate for their circumstances (Duncan, 2016; McKeon, 2015). Third, such participation tends to open up discussions beyond topics that would have been on the agenda otherwise, particularly by revealing interconnections across the food system. Public participants, especially those from social movements, speak from their lived experience rather than professional siloes. Thus they reveal systemic interconnections between hunger and institutions or systems that might be considered irrelevant otherwise. Fourth, the public considers and evaluates risks differently than technical experts and regulators, so its involvement is critical for effective policy-making about food systems, which often involves risk trade-offs (Ankeny, 2016; Slovic, 1987). Fifth, civil society participation allows recommendations and suggestions to be implemented more rapidly and smoothly because of public support and help with communication (Duncan, 2016; Scholte, 2001). Finally, civil society includes actors who can monitor the public sector for accountability (Scholte, 2001). This "watchdog" function is often attributed to civil society, but may be difficult to reconcile with the limited capacity and political space afforded civil society in various forums or with the simultaneous expectation that civil society participation will lead to "good decisions" (Steffek & Ferretti, 2009). While journalists and academicians can document violations of the social contract between governments and their people, impacts fall most heavily on civil society and thus they have the strongest case for demanding accountability to their needs (McKeon, 2015).

Despite these advantages, civil society participation in decision-making about hunger and food insecurity can be contentious or openly rejected for a number of reasons. In the points below, I am drawing primarily on my observations and experience with the CFS and ways that different governments respond to civil society there. First, some governments (or public sector agencies at various levels) have little respect for civil society, or fear that the latter will expose negligence or even illegal activities of the government in question. This is related to the watchdog function of civil society, which can be unwelcome when a government has actions to hide. Civil society may be perceived as a threat to governments with weak democracies or which are compromised through cooptation by other interests. Second, government representatives may see civil society as being inconsequential, since social movements and NGOs do not generally bring substantial financial assets to the table. Third, civil society, especially social movements, is not clearly understood by actors within hierarchical structures. For example, ways that decisions are made within the social movement and who is chosen to speak for the movement seem obscure to actors accustomed to having decisions passed down from a supervisor. Therefore, actors coming from hierarchical government agencies or businesses may not believe that voices of civil society actors are legitimate. Finally, some governments do not have mechanisms in place to channel feedback from civil society, either because they do not understand its value or do not want to hear it. Thus, the process of engagement can be unwieldy and disputed (Scholte, 2001).

Given the delicate and vital nature of civil society participation, it is important that effective means of engagement be understood. The four models presented in this case study use different processes of selecting representatives and incorporating civil society input, which may well affect their outcomes and their potential to lead to food system transformation. Duncan and Pascucci (2017, p335) introduced a useful framework that compares "isomorphic" and "polymorphic" alternative food networks (AFNs). They claim that

> when a network of AFNs mimics the dominant relations of the regime they may gain in terms of incrementally shifting the regime towards their goals, but are unlikely to support pathways that present alternatives to the dominant regime. On the other hand, networks of AFNs that are organised around polymorphic relations to the dominant regime struggle directly against the regime. If they persist and succeed, [then] . . . the alternative practices they promote stand to have a more pronounced impact in terms of radical change.

They suggest that structures that are more bureaucratic and rooted in market-based relations will be less transformative than those with more communitarian and democratic structures for decision-making. I will return to this distinction in the discussion section below.

Vermont's Farm to Plate Network (F2P)

Vermont is a small state in the northeastern US, one of the most rural and demographically homogenous of all the states. It has about 7,300 farms, and produces more than half of the milk consumed in the six New England states. Vermonters have an unusually strong commitment to buying local food, with more farmers' markets per capita than any other state and at least 94 per cent of schools sourcing local food (NASS, 2017; Vermont Agency of Agriculture, Food and Markets, 2016).

Vermont is a leader in development of a comprehensive food system plan (Sawyer, 2017). This collaborative effort started in 2009 with state legislation that authorized the Vermont Sustainable Jobs Fund (VSJF) to create a ten-year plan that would increase economic development in Vermont's farm and food sector, create jobs in the farm and food economy, and improve access to healthy local food for all Vermonters. Development of the plan involved over 1,200 farmers, food producers, technical assistance providers, and farm and food sector industry leaders from across the state. The Strategic Plan was launched in January of 2011 and has 25 goals including farm viability, reduced food insecurity, increased local consumption, and reduced environmental impacts. Each goal has regularly updated indicators that show how well the state is making progress (VSJF, 2017a).

VSJF organizes a Farm to Plate Summit every year that brings people together face-to-face for workshops and talks. Its staff support five Working Groups (e.g., Consumer Education & Marketing, Farmland Access & Stewardship); five Cross-cutting Teams (e.g., Health, Food Access); and at least 19 Task Forces that meet regularly to tackle the goals. A Steering Committee provides overall network governance and guides the implementation of the Strategic Plan. VSJF serves as coordinator of the initiative (the Network and its associated activities) and maintains a clear, informative website and newsletter (VSJF, 2017a).

F2P has not only increased communication by food system actors across the state through the Network, but it also has been instrumental in drafting and pushing through innovative legislation. For example, Vermont implemented a progressive state-wide food recovery and waste recycling law (Act 148) that mandates eliminating food waste in landfills by the year 2020, phased in over a few years. Membership in the Working Groups, Cross-cutting Teams, and Task Forces is voluntary; people engage if they have interest and ability. Leadership of these groups rotates regularly and is elected by members. People who participate are knowledgeable but not privileged as experts; this facilitates a leveling effect in which people are respected for their expertise and experience rather than title or position. VSJF publishes an annual report of the F2P initiative every year, listing accomplishments toward meeting the goals. It also includes results of a survey to find out whether the public supports its coordination of F2P. In 2016, VSJF reported that 85 per cent say the Network is helping to build new relationships and stronger relationships, 93 per cent say they value the networking opportunities, and 85 per cent say that VSJF is effective in coordinating the Network (VSJF, 2017b).

Analyzing the efficacy of F2P is somewhat difficult, since independent reports and peer-reviewed literature are just beginning to emerge. But one way to see how it compares with other state food plans in New England is through the New England Food System Planners Community of Practice (COP), sponsored by Food Solutions New England to enhance regional alignment (FSNE, 2017). Vermont was the first state in New England to create a comprehensive plan and other states tend to look to it for guidance. In my years of observation and participation, I have been struck many times by the degree of social capital among food system actors in Vermont and their ability to work together with very little friction. This extends to cordial relationships among civil society, government staff, and university personnel (who, as in many small states, may be interrelated). F2P is clearly a big contributor to this, by bringing people together regularly to work on shared goals. Based on a survey of F2P participants, Koliba et al. (2017) point to information sharing as a strong factor of success, with participants reporting that this is a stronger "value proposition" for their participation than project collaboration or resource sharing. That is, F2P's ability to facilitate communication among network members is central to why people stay involved or deepen their involvement from simply attending meetings to becoming part of the organizational structure.

Since participation in F2P Working Groups, Cross-cutting Teams, and the Annual Summits is voluntary, people "vote with their feet" and some groups have erratic participation. Even within strong groups, not all members are in accord on all issues. For example, discussion of the annual workplan for the Food Access Cross-cutting Team (of which I am part) was drawn out over several months because some people favour immediate visible actions to improve food access and some prefer to discuss the merits of various options at length before making a commitment to a specific action. This division can lead to frustration among group coordinators, who experience difficulty in getting the group to agree on a plan. It reflects very different approaches to social change, with some people advocating for more confrontational measures and some preferring to make incremental progress within an existing system, related to the distinction raised in Chapter 1 between food movements that are more "reformist" versus those that seek significant political and economic transformation of the entire food system. When people who advocate for different positions on this spectrum need to work together in a single committee, their differences of temperament or ideas about how to effect change may slow down progress.

Another source of tension in the Food Access Cross-cutting Team has been over the right to food and food sovereignty. Two members of the Team led an effort to have Vermont endorse the right to food, and created a rubric showing a continuum from hunger awareness through hunger relief to the right to food and food sovereignty. While other Team members, notably some working with anti-hunger organizations, supported the right to food and food sovereignty in principle, they objected to the implication of the rubric that actions based in the right to food and food sovereignty were "better" than hunger awareness and hunger relief. They were concerned that people who work in organizations providing charitable services might feel that their work was being dismissed as less significant, although the rubric creators insisted that their intention was to encourage people to consider what more they might be doing beyond their current foci, by looking at actions that characterize rights-based or food sovereignty approaches.

Vermont's relative success in carrying out a state food plan compared with other states may be due to many factors that make it unusual. It is small (population 623,657 in 2017 – smaller than many cities) and rural. According to the most recent US Population Census, 61.1 per cent of Vermont's population lived in rural areas in 2010, making it the second most rural state. It is quite homogeneous demographically: Vermont is one of the whitest states in the US, with 94.4 per cent of the population white according to 2016 American Community Survey data. There are few large agricultural industries that would advocate for private interests. Although dairy is important, the three largest milk buyers and processors in the state are member-owned cooperatives (Cabot, DFA, St. Albans), unlike states where the primary buyers of the main agricultural commodities are private or corporate. It has a deep history of countercultural "back-to-the-landers" who started farms in Vermont beginning in the 1960s and continue to colour local and state politics (Daloz, 2016; Kelley, 2016). Vermont has strong progressive political values: the legislature has been majority Democratic since the mid-1980s although the governorship has often been held by middle-of-the-road Republicans, and Vermont's two US Senators (Bernie Sanders and Patrick Leahy) are among the most consistently progressive Democrats in the country.

I worked on Maine's state food strategy between 2013 and 2015, and I observed much less cohesion and trust among food system actors there, despite Maine sharing many characteristics with Vermont: it vies with Vermont as being the most rural state, it is predominantly white, it has no large metropolitan areas, and it has strong support for local food. There are big differences in culture and politics between the two states, however, with Maine having much greater independent and libertarian constituencies and less of the countercultural brand that Vermont has created (Kaufman and Kaliner, 2011), plus a powerful and conservative potato industry. Perhaps the legacy of the back-to-the-landers in shaping a cooperative spirit ("we're all in this together") also contributes to Vermont's success. Regardless of specific causes, VSJF has been successful in building on collaborative attributes of the state to foster participation and buy-in to the F2P Strategic Plan.

To summarize, F2P is a successful collaboration in a progressive state, promoting food security in the context of economic growth and more jobs for the state of Vermont. While its structure allows people working on food security and nutrition to meet regularly and coordinate their efforts, F2P is unlikely to lead to a radical transformation of Vermont's food system. The cohesion of the group relies on social capital and compromise, so trust is high among members, perhaps at least in part because the group will not openly advocate for specific policies nor take controversial positions.

Committee on World Food Security (CFS)

The CFS was established as a United Nations (UN) Committee in 1974 in response to a global food crisis, in order to review and follow up food security policies (CFS, 2015). It was reformed in 2009 after an unanticipated global food price surge which drove 75 million additional people into severe food shortages and led to the overthrow of several State governments (Lagi et al, 2011; FAO, 2008). Member States and UN administrators realized that the CFS structure had not been adequate to meet its core goal of predicting a major disruption in food access, and changes were needed (IRIN, 2010). One of the changes made was allowing civil society to participate as a full member of the CFS by contributing to discussions on the floor, joining Open-ended Working Groups, organizing side-events, and helping to plan the annual CFS meetings. Government delegations are the only voting members, but otherwise civil society has a role equal to the private sector, State delegates, international scientific organizations, and representatives from other UN agencies that deal with food and agriculture. Civil society is self-organized through a Civil Society Mechanism (CSM) which meets for two days before the annual meetings of the CFS and debates issues, creating a strategy for how to address various points likely to come up in the CFS meeting. Through the year, members participate in CSM Working Groups or Open-ended Working Groups which also have representatives from the other bodies of the CFS (CFS, 2009).

The main product of the CFS prior to its reform was the Voluntary Guidelines to Support the Progressive Realization of the Right to Adequate Food in the Context of National Food Security (2005). Since reform, products include Principles for Responsible Investment in Agriculture and Food Systems; Voluntary Guidelines on the Responsible Governance of Tenure of Land, Fisheries and Forests in the Context of National Food Security; the Framework for Action for Food Security and Nutrition in Protracted Crises; and the Global Strategic Framework for Food Security and Nutrition, which records all major decisions by the CFS.[1] In addition, since 2009 the CFS has produced one or two evidence-based reports with policy recommendations each year on issues that influence or are influenced by food security, such as climate change, livestock production, and nutrition (CFS, 2017). The reports are overseen by a High Level Panel of Experts which includes civil society members, and produced by writing teams that always have civil society representation (Duncan & Barling, 2012).

In terms of volume of output, the CFS has made a huge leap since 2009; and moreover, participation by people from each representation group has increased each year (as is documented in summary reports from each year available through the "Products" website referenced in endnote 1). The reformed structure has been held up as a model of effective civil society participation for other UN forums, such as the Convention on Biological Diversity and the group that drafted the Sustainable Development Goals and Agenda (CFS, 2015; and sessions of CFS 2015 meeting). It is not an unqualified success from any perspective, however: the CSM regularly assesses wins and losses and believes it must constantly struggle to keep agreed-upon decisions in force, such as basing all products and policies in the right to food and ensuring that civil society is able to participate in planning major events. These ground-rules are part of the 2009 reform document and have been reiterated in the Global Strategic Framework, yet some Member States are eager to countermand them. A few Member State delegations have been obstreperous or resistant to CFS decisions supported by civil society and the majority of delegates, and have delayed progress on certain issues due to ideological reasons that are not supported by evidence. An innovative rights-based monitoring mechanism, specified as part of the 2009 reform, has been a particular sticking point for the US delegation, which repeatedly resisted efforts of civil society to implement a clear monitoring framework and process.

CSM is hardly unanimous as well; groups within often push for greater say, although from outside CSM the structure appears to be seamless. The CSM Forum during the weekend before the CFS meets occasionally and has heated debates among representatives of different constituencies. Yet disagreements do not spill over into the CFS week: CSM members value solidarity, so they support each other's positions whenever possible. One US government delegate commented to me that he could not understand how CSM holds together so tightly, because he was sure that there must be divisions within. This exemplifies the differences, mentioned above in discussion of reasons why civil society participation may not be seen as legitimate, between a hierarchical structure (government bureaucracies) and a level solidarity-based structure (the CSM). Ironically, the 2009 reform of CFS which allowed civil society representation as an equal was driven at least in part by a desire to increase the body's overall legitimacy (CFS, 2015) and the quality of its decision-making.

To summarize, the CFS is a remarkable and unique international food governance forum, focused on food security. No other intergovernmental forum allows civil society to participate as an equal with other representatives and government delegates. This inclusion of civil society in all aspects of planning, programming, discussion, and evaluation, particularly given the emphasis that the CSM places on social movements, allows those who are most vulnerable in the global food system to confront governments that are failing to uphold their obligations under the right to food and nutrition. The grounding of the CFS in the right to food and nutrition and rights-based approaches, including accountability and inclusive participation of all who are affected by its decisions, has potential for real transformation of the global food system in the direction of access to healthy food for all. Whether the CFS can realize this potential is uncertain, however. Although the CSM is a force for transformation, it is only one of several recognized entities in the CFS. A solid monitoring system showing whether countries are adhering to the guidelines created by the CFS, including the Voluntary Guidelines on the Progressive Realization of the Right to Food, is critical to demonstrate which countries are in compliance with agreements that their delegates make at the CFS, and to hold non-compliant countries accountable. There is always a tension within the CSM between the desire to point to non-compliance, and the fear that countries fingered in this way will simply withdraw from the CFS or become even more obstructive. So far, there have not been any withdrawals; but obstruction by governments to a transformative agenda is common.

Additional governance models that address food security at other scales

While my participation in food system governance for food security and nutrition has been primarily with the Vermont Farm to Plate Network and the Civil Society Mechanism of the Committee on World Food Security, I want to raise two other innovative governance models for comparison, in order to look at ways that civil society is engaging across different scales. Food policy councils (FPCs), addressed in greater detail in Chapter 8 of this book, typically consist of diverse stakeholders who work together to strengthen and expand the food environment within their communities. They may be created by the local government or by a public or private organization. The Johns Hopkins Food Policy Network has been surveying FPCs in the US and Canada each year since 2013 and reports 324 active groups in their most recent survey (Sussman & Bassarab, 2016). While 38 per cent are county-wide, some serve a city or municipality, tribal unit, or region. The structure of FPCs varies considerably: for example, only 15 per cent reported that government employees are members.

FPCs have proven to be effective entities for reviewing and recommending state and local food policies (Public Health Law Center, 2017). Although their goals vary widely, improving access to healthy food for people who are not well-served in the geographic area of the FPC, particularly through greater consumption of locally produced food, is nearly ubiquitous as a goal.

A recently developed FPC self-assessment tool asked participants to indicate their level of agreement with statements about their FPC's impact on the food system in which they work. Respondents reported that the greatest impacts were on facilitating changes in policy or practice that will promote the FPC's mission, increasing opportunities to purchase locally produced agricultural products, promoting social justice in the food system, and increasing access to healthy food in their community. They reported the lowest impact on promoting occupational safety and humane treatment of animals in the agricultural sector (Calancie et al, 2017). This study also noted that most FPC members are white women aged 35 to 65 years, and that greater racial diversity was associated with the FPC's ability to change public policy and improve community systems to prevent adverse health outcomes (Calancie et al, 2017).

The other governance model I want to introduce is the CONSEA in Brazil. Brazil has become an international exemplar for reducing hunger and food insecurity, and the CONSEA is part of its innovations. Brazil integrated the human right to food in its Constitution in 2010, enabling a legal framework that applies a rights-based approach to food security in Brazil. Civil society has played a decisive role in advancing the right to food in Brazil, particularly through the National Council on Food and Nutrition Security or CONSEA, a political and social advisory body that proposes guidelines for actions related to food security. CONSEA was originally created in 1993, interrupted its functions in 1995 and came back with renewed force in 2003 with the Zero Hunger Program. CONSEA is composed of two-thirds civil society and one-third government representatives, and chaired by a civil society representative, giving unusual dominance of civil society. It has bodies with similar make-up at many municipal, state, and federal levels; the different scales are integrated to achieve a national food security strategy and work together to ensure policy coherence (Strakos & Sanches, 2017). Coordination across ministries and dialogue with civil society at all levels are considered to be keys to the success of the Zero Hunger Program (OECD, 2016). Pérez-Escamilla et al. (2017, p. 70) in a discussion of the CONSEA conclude that "the inclusion of civil society stakeholders in decision making, oversight and resource allocation processes may indeed have beneficial rewards to FSG [food security governance]".

As a result of Brazil's Zero Hunger Program activities, 29 million people were lifted out of poverty between 2003 and 2014. Socio-economic inequality decreased 6.6 percentage points in the Gini coefficient in the same period (from 58.1 to 51.5). The income level of the poorest 40 per cent of the population rose, on average, 7.1 per cent in real terms, compared to a 4.4 per cent income growth for the population as a whole. And the proportion of people suffering from chronic undernourishment decreased from 7.5 per cent in 2008–2010 to 6.9 per cent in 2011–2013 (Strakos & Sanches, 2017). The Food Insecurity Experience Scale that the UN FAO developed to compare food insecurity across countries in a reliable way reported that Brazil's level of moderate food insecurity in 2014 was 8.3 per cent and severe food insecurity was less than 0.5 per cent, in contrast with neighbouring Ecuador (22.2 per cent and 8.7 per cent respectively) and Chile (12.0 per cent and 3.7 per cent). Other South American countries range between Chile's and Ecuador's figures. In contrast, FIES shows that moderate and severe food insecurity percentages in the US are 10.2 per cent and 1.2 per cent, and in Canada 8.0 per cent and 2.0 per cent (Cafiero et al., 2016).

In addition to operating at different scales than the CFS and F2P, the purpose and mission of the models introduced above differs. However, each of the four models discussed in this chapter has an explicit interest in eliminating or reducing hunger and food insecurity. Establishing and maintaining each structure require significant resources of time and money, so getting a better grasp on their respective strengths and challenges is important for understanding how to best reduce hunger and food insecurity.

Discussion

This section addresses four key factors that seem to facilitate or impede the effectiveness of the governance models introduced in this chapter and more specifically, civil society engagement in each model, and their ability to facilitate transformative food system change that would lead to all people having access to healthy food. In addition, I situate the models in this chapter on the continuum of governance discussed in Chapter 1.

Each of the governance models has accomplishments attributed to it, and perhaps as significantly, civil society actors continue to participate in them because they believe in their efficacy. If the organizations were taken over in some way by other interests and no longer served the public interest, one would expect civil society organizations to stop participating. In fact, that was the case with the CFS until the 2009 reform: La Via Campesina refused to engage, but was instrumental in setting up the terms of the reform and has been an influential participant since then (McKeon, 2015).

Role of civil society, especially the most marginalized people, in each model

Who is at the table in multi-actor forums and who is able to speak are very important. People who are poor, suffering from hunger, or in precarious political positions (e.g., because they are undocumented farm workers) often cannot attend frequent meetings, negotiate successfully for voice, use sometimes-obscure official language, or become well-versed in law and policy; so they will be under-represented unless strong efforts are made to involve them. Such efforts can include support for transportation, lodging, and mentoring, or a stipend. At the international scale, people who are marginalized from food systems policy-making and realizing the right to food include indigenous, peasants, ethnic minorities, and women; but the categories of "marginalized" varies by country and context. For example, race and indigeneity are especially important in North America.

Of the four structures included in this case study, only the CSM has made a distinction between social movements and NGOs. The former group has priority, must provide one of the leaders of each CSM Working Group, and is given most of the CSM speaking roles on the floor of the CFS. This is because social movements have direct experience with the circumstances of hunger and rights violation. Social movements demanded that they be allowed to speak for themselves a few years ago, after years of growing frustration with NGOs that wanted to "represent" their views. At present, NGO representatives and academics who participate in the CSM speak from their own expertise and experience; but if they consistently spoke against the interests of social movements such as La Via Campesina or the World Forum of Fisher People, their credibility would almost certainly shrink, as would their ability to fully participate in CSM processes with social movement actors. Participating NGOs and academics (including myself) generally do not perceive this to be any kind of censorship, but rather an opportunity to learn from social movements. There is wide recognition in the CSM that social movements have unique and deep expertise which must be made more visible in the CFS.

In contrast to the CFS with its explicit guidelines on civil society participation and self-organization, F2P has no guidelines or limits on how civil society engages: Working Groups and Cross-cutting Teams are open to anyone who chooses to attend meetings and tend to bring together people with interest and expertise in one facet of food system sustainability, such as increasing food access, training new business managers, or promoting land conservation. This may contribute to the relatively smooth operation of F2P Working Groups and their successes on policy and coordination across the state; people within each group are familiar with the issues. However, Working Groups can sometimes get "stuck", since the leadership has no way to push the group toward action other than persuasion and participants may have different ideas of how to proceed. But at its best, F2P holds a shared vision up to all participants, and encourages healthy and respectful dialogue and debate from which collective action proceeds.

The issue of whether people experiencing hunger and food insecurity should be participating in meetings of the Food Access Cross-cutting Team of F2P has come up recently, with some people asserting that their participation is essential to make sure that the group understands the lived experience of people in our state who are food insecure. However, others see the Cross-cutting Team as more of a higher-level communications mechanism among people who work directly in their professional capacities with poor and food-insecure people. The Team has not resolved this issue yet; but meanwhile, it has not yet made the effort to change the times and places it meets to make attendance of marginalized people more feasible, nor offered a stipend and daycare to those who would need this to attend meetings.

Beyond simple inclusion, is civil society respected for its unique contributions, or allowed to participate in governance mechanisms through tokenism or to greenwash food system decision-making structures? Vermont F2P and the reformed CFS were created as they exist explicitly because of civil society input: even though each was created through legislation or a formal decision among government representatives, the idea came from civil society and civil society helped to create the guidelines for whom should participate. This was also the case with the CONSEA, as it is now formulated. While the CSM is the place designated for civil society within the CFS, it is only one of six representation groups which have equal prerogatives; it does not dominate the CFS in any sense. Civil society participants make up the majority of and dominate F2P and the CONSEA. CONSEA's decision to have two-thirds of its membership from civil society is interesting; it is possible that a civil society majority is necessary to counterbalance the advantages of financial strength and political position of other actors in the food system.

FPCs vary tremendously both in their inclusion of civil society and in their efficacy. Some have disbanded because their members were no longer able to work together and adapt to changing circumstances, e.g., the Portland/Multnomah FPC, described by Coplin and Cuneo (2015), which dissolved in 2012 after ten years of operation. Since FPCs vary in their make-up, it is not possible to make generalizations about how civil society participates, although a FPC without any civil society representation would have little legitimacy. However, some of the earlier FPCs had very few civil society members. For example, the Connecticut State FPC (one of the first state-level FPCs) had members from all departments of state governments that had any conceivable connection with food, business representatives and farmers, but only one civil society representative (a person from an anti-hunger organization; Drake, 2010). The representative nature of most FPCs (e.g., one person representing environmental interests, one representing planning, one representing food access) means that organizations speaking on behalf of marginalized people, rather than marginalized people themselves, often fill the slots.

How different models function

There are real differences in the written or informal rules by which each model operates, with the commitment to transparency varying notably across models. Out of the structures examined here, F2P may be the most transparent: minutes are circulated after every meeting and Working Groups and contact addresses are available online. The documentation for indicators showing progress toward each goal is clearly displayed on the website. The CSM is transparent in posting all relevant documents online, including CSM perspectives and statements, membership, and coordinators of each CSM Working Group. But the CFS as a whole is much less transparent: reams of documents are issued before each CFS meeting and the outcomes of so-called "decision boxes" on each issue are included in each year's Global Strategic Framework; but unless people are in the room while Open-ended Working Group meetings are happening or decisions about policy recommendations from reports are hashed out, it is almost impossible to know which governmental delegations support and which obstruct each measure. The Private Sector Mechanism, the correlate of the CSM for business interests, is markedly less transparent than the CSM. Its membership and results of discussion are not available to the public.

How each model facilitates learning, especially gaining a systemic perspective

Food policy problems are often complex and driven by multiple interacting factors. The food system itself is multifunctional, simultaneously expected to meet goals ranging from producing adequate food to providing decent livelihoods for workers to protecting ecological integrity. A systemic approach is important to understand how the factors influence each other, where key points of intervention are, and how to calculate and balance trade-offs among different goals. Without this understanding, each sector or organization is likely to forge ahead in advocating for its own goals without awareness of how its actions might impede other sectors or organizations.

Bringing people with different perspectives and backgrounds together might facilitate learning, if this is an explicit aim of the multi-actor platform. But this is usually not enough unless participants have a genuine commitment to understand other perspectives. Bringing together people who have very different experiences, orientations, and working cultures may just accentuate their differences. Some of the structures examined here have mechanisms in place to help participants learn: for example, the reports from the CFS High Level Panel of Experts (HLPE) have this purpose. However, HLPE reports vary in their usefulness and how well they balance civil society, State, and private sector interests, depending on the members of the writing team. Even though every writing team includes civil society participants, they may be dominated by private sector or conservative government voices. Some of the reports (such as those on biofuels and livestock) were heavily influenced by the private sector and governments that are particularly attuned to private sector interests, in the opinion of people working in the CSM.

F2P has reports in the form of chapters in its Strategic Plan which deal with various food system issues, often produced and updated by academicians under contract to VSJF. These chapters are not the subject of deliberation in F2P Summits, however, so there is no good way to know whether they reflect civil society views well or how they influence debates. However, F2P posts regularly update indicators of progress based on the best and most authoritative research. The indicators were selected by F2P, and they contribute to shared learning about how and whether progress can be achieved on each goal.

Potential of models to lead to greater food system sustainability

How do these structures contribute to overall food system transitions to greater sustainability goals, beyond addressing hunger and food insecurity? Duncan and Pascucci's (2017) framework is relevant here, contrasting alternative food networks that are more bureaucratic and reformist versus those that are rooted in communitarian and democratic decision-making structures, and hence have greater potential for radical change. There are elements of bureaucracy and elements of communitarianism in each model examined in this case study; but the most bureaucratic structure is probably the CFS, and the least F2P. However, scale of operation and impact are important as well as structural type. A major transformation of the food system within the state of Vermont almost certainly would not be as impactful as a modest movement toward sustainability within the CFS, if CFS guidelines and principles were actually enacted at the country level. If we look at the CSM instead of the CFS as a whole, we find a strongly communitarian and democratic structure, based in solidarity and the shared belief that social movements must advocate for themselves rather than through mediators. This difference is part of the tension that CSM members experience, as they try to advance their agenda within the more bureaucratic CFS.

The ability to work across scales is another significant contributor to a model's potential to lead to greater food system sustainability. The CFS struggles with this aspect; one of its big challenges is ensuring that decisions made on the floor of the CFS meeting in Rome are translated into terms that people can understand at State and local levels, and then implemented. F2P manages the transition between state-level policies and programs to local implementation well, but it is important to remember that Vermont is very small.

The CONSEA excels in cross-scale work, since CONSEAs at different scales are nested up to the federal level. FPCs are non-scaleable by definition: they address a single geographic location. However, the Johns Hopkins Food Policy Council Network lists 12 US regional über-networks of FPCs and one in Canada. By connecting different FPCs, these organizations are trying to help them learn best practices from each other. The regional alignment group that brings New England state-level food system planners together, administered by Food Solutions New England, plays a similar role. In both of these instances, however, there is strong recognition that policies must be context- and place-specific and may need to be adapted if they are transferred to another location.

The priority given to marginalized people also weighs into a structure's potential to shift the food system toward greater justice and sustainability (McKeon, 2009). Such people have lost out disproportionately under capitalism and the neoliberal order that prevails in Western industrialized nations. For example, according to the Urban Institute, between 2007 and 2010 the average wealth of white families in the US fell by 11 per cent, but Black families saw their wealth fall by 31 per cent (as reported in Klein, 2017). The economic gains under neoliberalism have gone almost exclusively to the upper 1 or 0.1 per cent in the US (Sommeiller et al., 2016); and Oxfam International reports that three white men in the US have as much wealth as the entire bottom half of the population (Oxfam International, 2018). Indeed, capitalism rests on the theft of land from Indigenous people, the theft of natural resources from less-powerful countries, and the theft of people (slavery of Africans and Indigenous people) followed by merciless exploitation of labour. Therefore, justice demands putting marginalized people first. Reversing the momentum toward greater income inequality is essential for justice and food system transformation to be achieved, and racial equity is essential as part of this reversal.

None of the models examined here include anti-racism or radical income redistribution in their tenets. Therefore, their potential to lead to widespread justice and sustainability are limited; they are more likely to result in incremental progress. That said, some of the products of the CFS, such as the Voluntary Guidelines on Governance of Tenure, would result in far more equitable access to land if they were implemented. Given that CFS Guidelines are voluntary, however, such results seem to be more aspirational than achievable.

Finally, the extent to which models are grounded in human rights affects their potential to lead to transformative changes in food systems (HRBA

Portal, n.d.). The leading alternative framework to neoliberalism in the food system now is food sovereignty, which has gained proponents as the oppression of workers and small-scale producers has increased through practices such as land-grabbing, corporate consolidation, and the growing prevalence of precarious labor. Food sovereignty is strongest in the Global South, but most workers and small- to mid-scale producers in the Global North are also hurt by neoliberal policies (Fraser, 2016). However, food workers, producers, and consumers in the Global North have not joined forces in advocating for food sovereignty as yet.

Food sovereignty is based in the right to food, but it encompasses women's rights, peasants' rights, and the rights of people to control their own food system; the intersectionality of food sovereignty is a large part of its appeal (see Chapter 1). Key elements of public policies that strengthen food sovereignty include a right-to-food approach with accountability and decision-making put into the hands of those most affected by food policy decisions, alliances between peasants and other social classes and constituencies to negotiate policies with the state, and alignment across all scales of policy-making from local to global (Kay et al, 2018).

None of the models described in this chapter explicitly calls for food sovereignty; but in addition to mandated grounding of all decisions and guidelines made by the CFS in the right to food, women's rights have received increasing attention in its annual meetings. The Food Access Cross-cutting Team of F2P is embarking on a campaign to promote the right to food in Vermont, but this is somewhat contentious, as described earlier. The Brazilian CONSEAs fully support the right to food, which is recognized in the Brazilian Constitution. While individual FPCs may include the right to food in their mission statements, this is not well recognized in the US. The US is the only industrialized country that does not support the right to adequate food and nutrition; and US citizens are ignorant, for the most part, about what this right entails (Anderson, 2013). Although the right to food is an important underpinning of food sovereignty and quite important in framing genuine solutions to hunger and food insecurity, its recognition alone is not sufficient. It must be accompanied by significant structural reforms and coordination across different organizations and sectors of society (Pérez-Escamilla et al, 2017).

Connections with the continuum of governance in Chapter 1

Chapter 1 of this volume introduced a continuum of governance from multistakeholderism to polycentric models. Of the models mentioned in this chapter, the CFS rests squarely in the multi-stakeholder mode. F2P is closer to co-governance, given the close collaboration between state agency staff and civil society in design and implementation of policy. FPCs are at the multistakeholder end of the spectrum, although they are quite variable. Some may operate in a co-governance mode, if power-sharing mechanisms are in place to prevent dominance by state or business interests. The CONSEA comes closest to being dominated by public-interest concerns, given its majority civil society participants. It also functions more as a polycentric model through its nested municipal to federal councils.

Conclusion

Several of the attributes of the four governance models described in this case study stand out as enabling civil society actors to contribute to reductions in hunger and food insecurity. Perhaps most important is the role of civil society and social movements within civil society. Without strong representation in multi-actor coalitions, perhaps even dominance (as is the case in CONSEA), the voices of civil society are likely to be drowned by those who are better financed and politically connected. When the differing status and capacities of civil society actors are accommodated (for example, through stipends or assistance in being heard), they can be most effective. When those who are most vulnerable to harm through food system policies or have been harmed in the past by food system policies are given priority over those whose lives and livelihoods are not directly affected, better decisions ensue: the decisions made by the organization are more likely to address the causes and consequences of hunger and food insecurity. Civil society is best able to participate if consulted early, in the formation of a multi-actor coalition, regarding its structure and rules. If civil society is added later simply to add legitimacy to the coalition, frustration and misunderstandings are likely.

In terms of the function of a multi-actor coalition, transparency regarding membership and actions is extremely important to maintain trust. Transparency may require careful documentation of all proceedings, and communication in different media. Competent leaders are also critical; they should support the coalition from below, but not dominate interaction and discourse. This is the function of the "backbone organization" in Kania and Kramer's (2011) collective impact framework. Another aspect of collective impact is an emphasis on learning and measurable evidence-based indicators of progress. The latter must have civil society buy-in to be effective; a frequent complaint of civil society is that indicators are dry and technocratic, rather than reflecting the lived experience of people suffering from hunger and food insecurity. This point illustrates the importance of communitarian and democratic principles in the functioning of the organization; civil society tends to adhere to these principles much more than to bureaucratic or market-based principles. In fact, they often form in resistance to excessive market orientation and bureaucracy. A final point on function is the importance of integration across scales. Civil society engages best if the decisions and activities of multi-actor coalitions can be picked up and carried forward by similar structures at higher or lower scales.

While these characteristics can enhance the effectiveness of civil society participation in organizations attempting to eliminate hunger and food insecurity, a transformation of the food system that allows this goal to be achieved is unlikely without a more radical agenda that includes anti-racism; income redistribution; and the respect, protection, and fulfillment of human rights. Although each of the models considered here has its drawbacks and unique challenges, the CFS and CONSEA have some advantages in this respect because both are based in the right to food. Explicit recognition of the right to food and structural modifications in alignment with rights-based approaches could encourage more transformative action toward sustainability within the Vermont F2P and FPCs.

Acknowledgements

I would like to thank the social movement and NGO actors with whom I have been privileged to work in the CFS and Vermont's Farm to Plate Network. I have learned a tremendous amount from them, and hope that I have represented their views faithfully.

Note

1 See repository of major products at www.fao.org/cfs/home/products/en/

References

Anderson, M. D. (2013) Beyond food security to realizing food rights in the US. *Journal of Rural Studies*, 29, 113–122.

Ankeny R. A. (2016) Inviting everyone to the table: strategies for more effective and legitimate food policy via deliberative approaches. *Journal of Social Philosophy*, 47(1), 10–24.

Cafiero, C., Nord, M., Viviani, S., Del Grossi, M. E., Ballard, T., Kepple, A., Miller, M., & Nwosu, C. (2016) *Voices of the hungry: methods for estimating comparable prevalence rates of food insecurity experienced by adults throughout the world*. Technical Report Number 1. Rome, Food and Agriculture Organization. Available atwww.fao.org/3/c-i4830e.pdf

Calancie, L., Allen, N. E., Weiner, B. J., Ng, S. W., Ward, D. S., & Ammerman, A. (2017) Food Policy Council self-assessment tool: development, testing, and results. *Prev Chronic Disease*, 14(160281), E20.

CFS (Committee on World Food Security) (2009) *Reform of the committee on world food security final version*. Available at: www.fao.org/tempref/docrep/fao/meeting/018/k7197e.pdf

CFS (2015) *Committee on world food security*. Available at: https://sustainabledevelopment.un.org/content/documents/15195Session%2012_Committee%20on%20World%20Food%20Security%20(CFS%20Brief)_May_2015.pdf

CFS (2017) *Major products*. Available at: www.fao.org/cfs/home/products/en/

Coplin, A. K. & Cuneo, M. (2015) Dissolved: lessons learned from the Portland/Multnomah food policy council. *Journal of Agriculture, Food Systems and Community Development*, 5(2), 91–107.

Daloz, K. (2016) *We are as gods: back to the land in the 1970s on the quest for a new America*. New York, Public Affairs.

Drake, L. (2010) *Connecticut Food Policy Council*. Presentation for the CT SustiNet Health Partnership Child & Adult Obesity Task Force, February 5. Available at: www.ct.gov/sustinet/lib/sustinet/taskforces/obesitytaskforce/connecticut2010_linda_drake.pdf

Duncan, J. (2016) Governing in a post-political era: civil society participation for improved food security governance. *Advances in Food Security and Sustainability*, 1, 137–161.

Duncan, J. & Barling, D. (2012) Renewal through participation in global food security governance: implementing the international food security and nutrition Civil Society Mechanism to the Committee on World Food Security. *International Journal of the Sociology of Agriculture and Food*, 19(2), 143–161.

Duncan, J. & Pascucci, S. (2017) Mapping the organisational forms of networks of alternative food networks: implications for transition. *Sociologia Ruralis*, 57(3), 316–339.

FAO (Food and Agriculture Organization of the United Nations) (2008) *Hunger on the rise*. Available at: www.fao.org/newsroom/en/news/2008/1000923/

Fraser, A. (2016) *Global foodscapes: oppression and resistance in the life of food*. London, Routledge.

FSIN (Food Security Information Network) (2017) *Global report on food crises 2017*. Available at: www.fsincop.net/resource-centre/detail/en/c/877403/

FSNE (Food Solutions New England) (2017) *Regional alignment*. Available at: www.foodsolutionsne.org/six-states-one-region/regional-alignment

HRBA (Human Rights-based Approach) Portal (n.d.) *What value does a human rights-based approach add to development?* Available at: https://hrbaportal.org/faq/what-value-does-a-human-rights-based-approach-add-to-development

IRIN (Integrated Regional Information Networks) (2010) *Global: please manage the food better*. Available at: www.irinnews.org/news/2010/06/24/please-manage-food-better

Kania, J. & Kramer, M. (2011) Collective impact. *Stanford Social Innovation Review*, Winter, 36–41.

Kay, S., Mattheisen, E., McKeon, N., De Meo, P., & Moragues Faus, A. (2018) *Public policies for food sovereignty*. Hands on the Land, FIAN, Terra Nuova and TNI, Amsterdam, Rome and Heidelberg.

Kaufman, J. & Kaliner, M. E. (2011) The re-accomplishment of place in twentieth century Vermont and New Hampshire: history repeats itself, until it doesn't. *Theory and Society*, (40)2, 119–154.

Kelley, K. J. (2016) Historical society explores impact of back-to-the-landers on Vermont. *Seven Days*. Available at: www.sevendaysvt.com/vermont/historical-society-explores-impact-of-back-to-the-landers-on-vermont/Content?oid=3632371

Klein, N. (2017) *No is not enough: resisting Trump's shock politics and winning the world we need*. Chicago, Haymarket Books.

Koliba, C., Wiltshire, S., Scheinert, S., Turner, D., Zia, A., & Campbell, E. (2017) The critical role of information sharing to the value proposition of a food systems network. *Public Management Review*, 19(3), 284–304.

Lagi, M., Bertrand, K. Z., & Bar-Yam, Y. (2011) *The food crises and political instability in North Africa and the Middle East*. Available at: https://ssrn.com/abstract=1910031 or http://dx.doi.org/10.2139/ssrn.1910031

Marsden, T., Lee, R., Flynn, A., & Thankappan, S. (2010) *The new regulation and governance of food: beyond the food crises?* London, Taylor & Francis.

McKeon, N. (2009) *The United Nations and civil society: legitimating global governance – whose voice?* London, Zed Books.

McKeon, N. (2015) *Food security governance: empowering communities, regulating corporations.* London, Routledge.

NASS (National Agricultural Statistics Service) (2017) *State agriculture overview: Vermont.* Available at: www.nass.usda.gov/Quick_Stats/Ag_Overview/stateOverview.php?state=VERMONT

OECD (Organization for Economic Cooperation and Development) (2016) *Better policies for sustainable development 2016: a new framework for policy coherence.* Paris, OECD Publishing.

Oxfam International (2018) *Reward work, not wealth.* Oxfam Briefing Paper. Available at: www.oxfam.org/en/research/reward-work-not-wealth

Pérez-Escamilla, R., Shamah-Levy, T., & Candel, J. (2017) Food security governance in Latin America: principles and the way forward. *Global Food Security*, 14, 68–72.

Public Health Law Center (2017) *Food Council by-Laws.* Kansas Health Foundation. Available at: www.publichealthlawcenter.org/sites/default/files/resources/Kansas-Food-Council-Bylaws-2017.pdf

Sawyer, S. (2017) Food system lessons from Vermont. In: Lerch, D. (ed.) *The community resilience reader: essential resources for an era of upheaval.* Washington, Island Press, pp. 227–245.

Scholte, J. A. (2001) *Civil society and democracy in global governance.* CSGR Working Paper, 65/01, University of Warwick.

Slovic, P. (1987) Perception of risk. *Science*, 236, 280–285.

Sommeiller, E., Price, M., & Wazeter, E. (2016) *Income inequality in the U.S. by state, metropolitan area, and county.* Economic Policy Institute. Available at: www.epi.org/publication/income-inequality-in-the-us/

Steffek, J. & Ferretti, M. P. (2009) Accountability or "good decisions"? The competing goals of civil society participation in international governance. *Global Society*, 23(1), 37–57.

Strakos, P. F. & Sanches, M. B. B. (2017) State's international responsibility for the human right to food: implementation in Brazil through agroecology. *Revista de Dereito Internacional* (Brazilian Journal of International Law), 14(1), 35–53.

Sussman, L. & Bassarab, K. (2016) *Food Policy Council report 2016.* Johns Hopkins Center for a Livable Future, Food Policy Networks. Available at: https://assets.jhsph.edu/clf/mod_clfResource/doc/FPC%20Report%202016_Final.pdf

Vermont Agency of Agriculture, Food and Markets (2016) *Agriculture in Vermont highlights.* Available at: http://agriculture.vermont.gov/sites/ag/files/pdf/news_media/VT%20Ag%20%26%20Agency%20Overview%20Final%202016.pdf

VSJF (Vermont Sustainable Jobs Fund) (2017a) *Farm to Plate strategic plan.* Available at: www.vtfarmtoplate.com/plan/

VSJF (2017b) *Farm to Plate annual report, year 6, fiscal year 2016.* Available at: www.vtfarmtoplate.com/uploads/Farm%20to%20Plate%202016%20Annual%20Report_FINAL.pdf

7 Indigenous self-determination and food sovereignty through fisheries governance in the Great Lakes region

Kristen Lowitt, Charles Z. Levkoe, Ryan Lauzon, Kathleen Ryan, and Chief Dean Sayers

Introduction

Access to traditional or "country" foods is a key element of food sovereignty for Indigenous communities in Canada (Morrison, 2011; People's Food Policy, 2011). The importance of traditional foods is not only that they are nutrient-dense, but they also contribute to "cultural food security" due to their central role in maintaining identity, health, and survival (Martin & Amos, 2016; Power, 2008, p. 95). Throughout the Great Lakes region, hunting and harvesting of wild and traditional foods have long been essential to the sustenance and ways of life for First Nations and Métis people[1] (Hudson & Ziegler, 2014; MacCrimmon, 2002; Lytwyn, 1990). Since time immemorial, Indigenous people have lived on the land, making use of the abundant fish and wildlife. Fishing is of special importance to the Anishinaabe people around the Great Lakes, with the lifestyles, rituals, and daily practices surrounding fishing deeply integrated into the cultures of communities (Kuhnlein & Humphries, 2017; Tobias & Richmond, 2014).

With the arrival of European settlers in the 17th century, Indigenous fishing activities were forcibly disrupted. As a result of settler-colonialism, the centralization of power, a broken treaty process, and an imposed reserve system, Indigenous control of land and watersheds and access to fisheries were significantly restricted. In contrast to Indigenous governance systems rooted in traditional ecological knowledge and reciprocity with nature (Coulthard, 2014; Borrows, 2010), regulations established by the settler state take a narrow, technocratic perspective. This "resourcist" view treats fish as a resource that can be managed for efficiency and profit (Olson et al, 2014; Berkes, 2010). It also places humans in a dominant position over nature, as manifest in expert-driven and command and control style structures (Berkes, 2010). This is a fundamentally different perspective and relationship with nature than that held by Indigenous people (King, 2014; Koenig, 2005; MacCrimmon, 2002).

Indigenous people around the Great Lakes have resisted state control and, over time, established varying forms of self-governance and engagement with the settler state. Today, there are 75 First Nations around the Great Lakes in Canada, all involved in fishing activities to differing degrees and with varying levels of authority over their fisheries (Assembly of First Nations, n.d.). In this chapter, we present two case studies of Indigenous fisheries governance systems that both exhibit models for self-determination[2] and a commitment to traditional ecological knowledge, albeit with different approaches to engagement with the settler state. Drawing on a review of key documents and interviews with decision-makers in each community, we explore the evolving governance arrangements of Batchewana First Nation of the Ojibways (BFN) located on the eastern shores of Lake Superior and the Saugeen Ojibway Nation (SON) on Lake Huron. We describe these communities' different governance arrangements and aspirations, strategies used to exercise jurisdiction, and perceptions of opportunities and limitations in organizing towards enhanced equity for sustainable fisheries. Ultimately, we seek to contribute to a sharing of governance experiences to enhance the potential for Nation–to–Nation relationship-building and support Indigenous movements towards self-determination and food sovereignty.

Working as a co-author team, we present a community-based perspective which is lacking within the Great Lakes fisheries literature. The author group, consisting of two academics, one staff member from the Saugeen Ojibway Nation, one staff member from the Chippewas of Nawash Unceded First Nation, and the Chief of Batchewana First Nation worked together to collect stories, develop the case study descriptions, and ensure that the stories told resonate with the communities perspective. The First Nations ethical principles of Ownership, Control, Access, and Possession (OCAP) guided all aspects of our research collaboration. The final version of the chapter was approved by each community prior to publication.

We begin by providing some context of Indigenous fisheries governance in the Great Lakes. We then turn to our case studies, describing each community's approach to fisheries and their governance arrangements with the settler state. In our discussion, we consider the case studies in relation to the governance engagement continuum discussed in Chapter 1 of this book and offer reflections and insights for the food sovereignty movement in Canada.

Historical context for fisheries governance

In the late 1700s to mid-1800s treaties were signed between Indigenous people and the British and Canadian governments (the settler state). The result of the treaty process was that Indigenous communities' fishing activities were restricted to reserves[3] (MacCrimmon, 2002; Lytwyn, 1990). Settlement of the Great Lakes region following the signing of treaties further restricted resource use by Indigenous people as settler commercial fisheries developed and out-competed Indigenous and subsistence fisheries (Brendan et al, 2012). In 1857, the Canada Fisheries Act was established to manage the settler commercial fisheries by establishing regulations to protect fish stocks (Brendan et al, 2012). However, the Fisheries Act did not recognize Indigenous fishing rights; as a result, Indigenous fishing activities became subject to licensing requirements, closed seasons, and other measures, and were restricted to fishing for subsistence (Brendan et al, 2012), representing a significant diminution of Indigenous rights (Koenig, 2005; Blair, 1997). As LeBlanc and Burnett (2017) argue, "the basis of all Indian legislation has been to force Indigenous people to adopt Euro-Canadian forms of governance, private property, individualism, and nuclear patriarchal families" (p. 25).

Indigenous people have continued to resist these efforts. An important development was the recognition and affirmation of Aboriginal and Treaty Rights in section 35(1) of the Constitution Act, 1982. However, there remain different views on the nature, scope, and content of these rights. Today, much of what happens in respect to Indigenous rights depends on decisions from the Canadian courts (Christie, 2007). Since 1982, legal decisions have clarified the nature of these rights in relation to natural resources and fisheries. For example, *Sparrow* and *Marshall* were foundational Supreme Court of Canada decisions. While not rejecting government authority in respect to resource conservation, they affirmed that federal or provincial regulations must not infringe in a major way on Indigenous rights or discriminate against Indigenous fisheries (Brendan et al, 2012; Gaden et al, 2012).

Central to the exercise of Aboriginal and Treaty Rights is the premise that along with the right to fish comes the responsibility to protect and govern resources (Brendan et al, 2012). It is in the interests of Indigenous people to govern their own fishing resources, as the values underlying resource use differ from those of settler governments (Brendan et al, 2012). MacCrimmon (2002) writes, "State governments manage fish resources based on the greatest potential economic return to the state, whereas the [Tribal] bands use of the resource is based on food, tradition and livelihood" (p. 251). These values and worldviews underlying Indigenous relationships to land and water have been well documented (Coulthard, 2014; Simpson, 2011; Borrows, 2010). Coulthard (2014) describes how Indigenous people understand land as a "system of reciprocal relations and obligations [that] can teach us about living our lives in relation to one another and [the] natural world in nondominating and nonexploitative terms" (p. 13). These ideals relate to the concept of food sovereignty, a set of theories and practices grounded in the knowledges of peasants, Indigenous people, pastoralists, and fisherfolk that emphasize the democratization of food systems and the rights and autonomy of food producers and harvesters (Patel, 2009; Nyéléni Declaration, 2007). As one of the world's largest social movements, food sovereignty takes aim at state and corporate control of the food system to assert that food is a basic human right and that governance must be grounded in the needs and desires of communities in relation to the health of ecosystems (Desmarais et al, 2017). At the forefront of Indigenous food sovereignty is the recognition that food is core to self-determination and a sacred part of maintaining relationships with land, plants, and animals (Morrison, 2011; People's Food Policy, 2011).

In this context, many Indigenous communities across Canada are seeking to redefine existing institutions of power and authority to exercise their governance of land and water (Natcher, 2001). Some communities are utilizing co-governance frameworks as a means of working towards self-governance, in terms of determining their own affairs according to their established cultural and legal institutions (Christie, 2007). In many cases, co-governance frameworks have been put in place through comprehensive land claim agreements requiring the formation of joint resource management boards (comprised of settler government agencies and Indigenous community representatives) to oversee the management of natural resources in settlement areas (Natcher, 2001). In other instances, co-governance agreements have been established in situations of resource "crisis", often involving migratory species and disputes over jurisdictional boundaries and authority (Natcher, 2001). However, not all Indigenous communities have entered co-governance frameworks, particularly where they believe these frameworks may not recognize the full extent of their inherent rights[4] and jurisdiction, or where they may fundamentally conflict with their values or self-governance aspirations. In either case, the aim of co-governance and self-governance efforts, as Christie (2007) explains, is that "the Crown recognize something they [Indigenous nations] already possess – that they have always possessed – the exercise of which has been blocked by the Crown for many generations" (p. 8).[5]

In the realm of food policy, LeBlanc and Burnett (2017) note that Indigenous people's exclusion occurs predominantly in the natural resource management sector at the provincial level. In terms of Great Lakes fisheries, under the Canadian settler-state regulatory regime, primary authority for fisheries management resides with the Ontario Ministry of Natural Resources and Forestry (OMNRF),[6] including responsibilities for management practices, quota allocation to each type of fishery (i.e. Aboriginal, recreational, commercial, and baitfish), enforcement, and fish culturing and stocking (Boudreau & Fanning, 2016).

OMNRF focuses on Western science as a basis for decision-making (Boudreau & Fanning, 2016). However, OMNRF's 2015 Ontario Fish Strategy refers to Indigenous rights and knowledge. Specifically, the Strategy acknowledges that Indigenous fisheries pre-date the Province of Ontario, that they are important to diets, culture, and economies, and that OMNRF is trying to find ways to incorporate traditional ecological knowledge into management (Government of Ontario, 2015). According to the Strategy, Indigenous people do not need a license if they are fishing within their traditional territories for food, social, or ceremonial purposes. However, Indigenous commercial fisheries still receive quota allocations in accordance with OMNRF's interpretation of case law, which provides allocation first to conservation and second to rights-based fisheries. OMNRF's main approach has been to negotiate arrangements with Indigenous communities for commercial fisheries. Since 1994, in addition to standard commercial fishing licenses, the OMNRF's preferred approach is to issue Aboriginal Communal Fishing Licenses. While these arrangements bring some decision-making power to First Nations, they also function to retain ultimate control with the settler state. Currently, there is a patchwork of different governance arrangements in place between OMNRF and First Nations across Ontario, including communities that have negotiated terms of license conditions with OMNRF and others that have not. To explore two different arrangements, we now turn to our case studies.

Batchewana First Nation of the Ojibways

Batchewana First Nation (BFN) of the Ojibways is signatory to the Robinson Huron Treaty (1850), with traditional lands stretching from Whitefish Island to Batchewana Bay (see Figure 7.1). Through the Pennfether Treaty (1859), BFN ceded much of their territory to the Crown in return for the unkept promise of land and the redistribution of interest to band members on an annual basis (Syrette et al, 1977). Over time, Batchewana has reclaimed parts of its territory through purchase and land claims. Today the population is about 2,400 people, of which approximately 72 per cent live off-reserve. Colonization has continued to disrupt longstanding practices and ways of life, including a reduction in the ability to undertake both subsistence and commercial fishing activities on which they had relied for centuries (Tobias & Richmond, 2014). This has included imposed settler state regulation as well as limited resources to purchase and maintain equipment and supplies. Nevertheless, BFN members have continued to fish and assert their inherent Aboriginal and Treaty rights.

Today, BFN operates one of the largest fisheries on the Canadian side of Lake Superior. Twenty-seven captains and their crew, family members, and First Nations staff are involved in the fishery. Many fishers and their households depend on their catch for sustenance and as a vital component of livelihoods. Fish not consumed within the community are sold to local processing plants, regional restaurants and retail outlets, and through a busy farmers' market in Sault Ste. Marie.

Different from the provincially managed system based on ideals of conservation and recreation (OMNRF, 2015), BFN have maintained a fisheries management system rooted in traditional ecological knowledge and developed in alignment with oral teachings and responsibilities. This is described as a governance system that works with nature to regulate the fisheries. These approaches are driven by cultural identity and the intergenerational transfer of knowledge built over time between the people and their relationships to the natural environment. Chief Sayers, BFN's elected and hereditary Chief explained this approach to fishing:

> Historically, we would move with the different times of the year with what was happening on the earth. The elders tell me that in the spring, when the poplar leaves are the size of a thumbnail, we would be in a certain place. When the strawberries were ripe, we would be in another place. And when the raspberries were ready, we were in another place. When the yellow scum on the edge of the water appears, we would be in another place. And we kept moving depending [on] what's happening

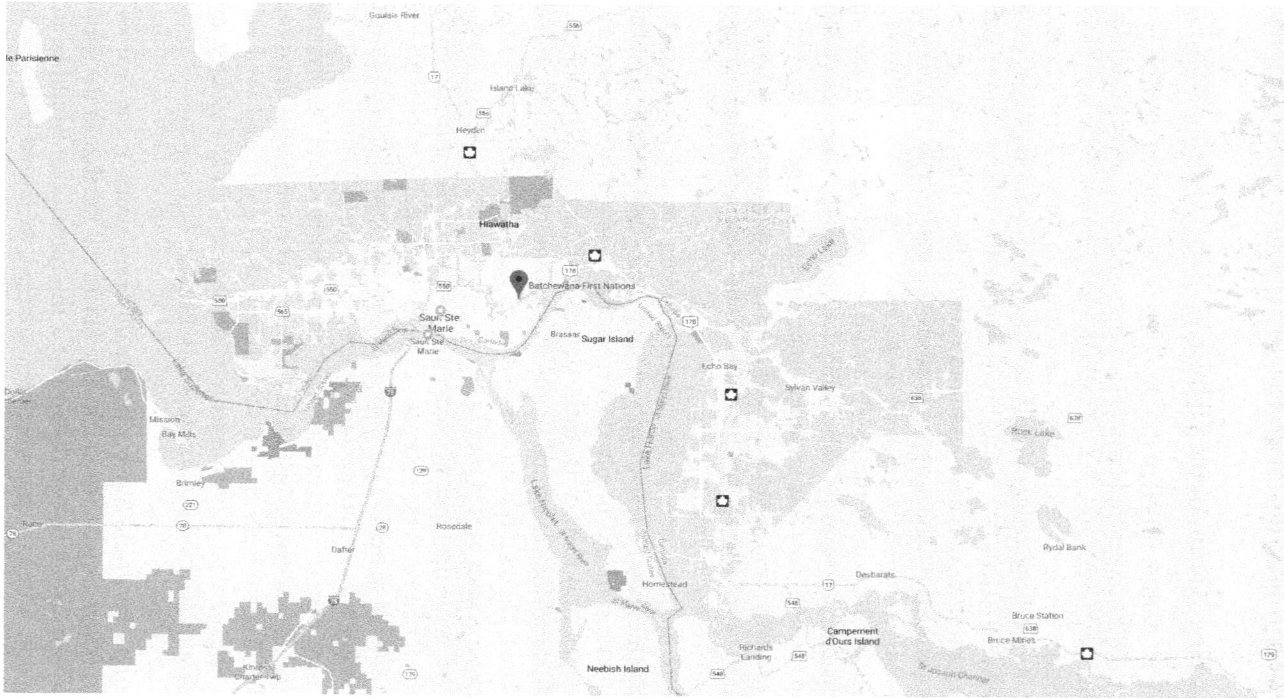

Figure 7.1 Location of Batchewana First Nation (Rankin Reserve)

Source: Map data ©2018 Google

> with the earth all along the eastern shore of Lake Superior. Even today they fish that way. They fish with the weather, they fish with the seasons, they fish with what's happening with the environment. This is the traditional ecological knowledge and environmental knowledge that our people still have today – through exercising the unextinguished right to manage the fishery, absent of any of the mainstream government approaches.

Chief Sayers is adamant that BFN retains sovereignty over their lands and waters along with governance over the fisheries: "We are the first level of law in these lands. Not the Ontario government, nor the Canadian government". He explained that, for a long time, OMNRF restricted BFN's access to commercial licenses and severely limited the community's livelihood. Claiming their Aboriginal and Treaty rights, BFN rejected those rules and reasserted their historic fishing rights. He explained:

> When we rejected fishing licenses from Ontario, we actually issued our own based on the treaty saying we had the right and we didn't need a license. We could fish and hunt as we were in the habit of doing based on our self-determination, not on a foreign government telling us what we can do.

This claim for the right to manage their own fishery and fish commercially was upheld by a court case in 1988 [*R. v. Agawa*, 1988 CanLII 148 (ON CA)]. In response, BFN has set up their own fishing authority separate from both the provincial and federal governments. They have continued to actively assert their sovereignty and are wary of entering management relationships orchestrated by the state. A major concern is that when sitting at the table with government representatives, scientists, anglers' associations, and environmental groups, BFN becomes just one voice. Chief Sayers explained:

> If there is eight of them at that table, we sit and we listen. Let's say the anglers would like to see the introduction of salmon into Lake Superior. They will say it employs a lot of the resorts, brings in tourism, and it will bring in a lot of money for everybody. But we say we don't agree. It's an invasive species. They are carnivores and will eat all of the other small fish and eggs . . . So, they'll call a vote and six or seven of those organizations will vote in favour of it and maybe one or two tribes will say they are opposed. Because we are at that table, our vote is watered down and what happens does not align with our inherent obligation to protect the Lake. So, we won't give it power because it is an illegal assertion of jurisdiction. They don't have the jurisdiction to manage the fishery. There is no treaty, at least from our perspective, that gave the fishery jurisdiction to the settler governments and their governance mechanisms. So, we cannot sit there or we recognize their theft and illegal jurisdictional assertions.

Maintaining a management system that is different than the one mandated by the state has led to many tensions. To date, OMNRF has not formally recognized BFN's jurisdiction and traditional management plan. Beyond a lack of recognition, there have been significant tensions based on fundamental differences in approaches to governance. A scientist with OMNRF suggested that the Ministry wants to enter into greater collaboration with BFN, but that their attempts have been met with resistance. He explained:

> It makes sense for us to work together and share information and develop a rapport and some mutual respect and develop a relationship. That's what we'd like to work towards but we haven't been able to have anything near that type of conversation with Batchewana because they're much more rebellious or, I don't know, aggressive.

Another scientist with OMNRF spoke about BFN overfishing, fishing outside their territory, abandoning nets, and ignoring the conservation-based management protocol. Both scientists also noted that BFN do not contribute any information about their catch or additional data needed. A former OMNRF employee who was part of negotiations with BFN to enter into a governance relationship noted the importance of a system based on underlying OMNRF rules:

> Yes, you have a right to a fishery, we're going to give you that but there has to be a license, there has to be an overall umbrella control. You can't just say everybody go fishing, period. You can't have a system like that. It would endanger the fishery.

Recognizing the epistemological differences in the approaches to governance, Chief Sayers explained:

> Their judges swore allegiance to the crown and the Crown's laws. So, if our laws contradict that, they are going to follow their laws and they'll charge us and find us guilty of offense against their law, even though they created it based on corporate and industrial lobbying. That's what their law is. It doesn't matter how logical you are if it breaks that law. The cards are stacked against us.

Chief Sayers maintains that BFN aims to "reclaim our governance rights or unextinguished jurisdictional rights and exercise our inherent obligations" through governing the fisheries:

> We recognize what needs to be done in order to bring the Lake back to its healthy status. It should be done in a ceremony. It should be done in a way that connects with our ancestors. If we don't know something we can find out spiritually what that answer is and how we can incorporate that into our models today.

Saugeen Ojibway Nation

The Saugeen Ojibway Nation's (SON) traditional territory consists of 6,500 km^2 of land extending from the Saugeen (Bruce) Peninsula, south to the Maitland River, and east to the Nottawasaga River. The territory includes numerous tributaries, over 500 km of shoreline and 10,000 km2 of Lake Huron (LaRiviere & Crawford, 2014; see Figure 7.2). SON has a 930-hectare hunting reserve on the northern Saugeen Peninsula, an Aboriginal Title claim to the waters and lake bed of a large extent of Lake Huron and Georgian Bay, and exclusive commercial fishing rights negotiated with the Crown that cover most of their traditional waters around the Saugeen Peninsula in Lake Huron and Georgian Bay.

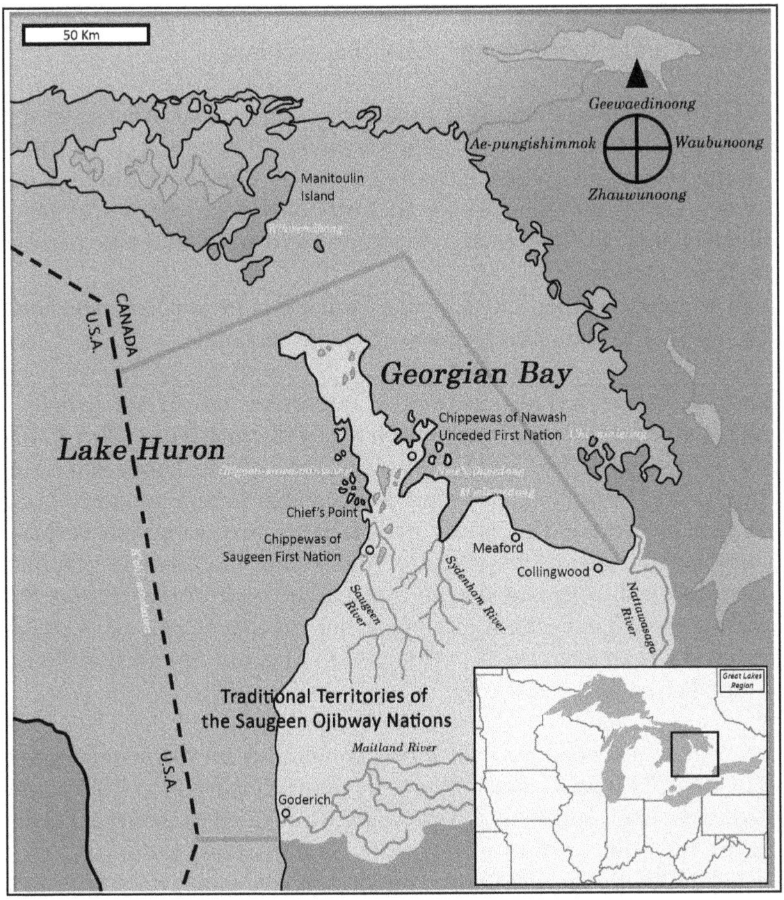

Figure 7.2 Map of SON's traditional territories

Source: Reprinted from Lowitt et al (2018), with permission from Elsevier

SON is the collective name for two sister communities that share this traditional territory: the Chippewas of Nawash Unceded First Nation and the Chippewas of Saugeen First Nation. The two communities are situated approximately 60 km apart on the Saugeen Peninsula, each with an on-reserve population of about 1,500 people. A Joint Council, consisting of elected councilors from each of the communities, provides a shared governance structure for the territory. From time immemorial, the Saugeen Ojibway have engaged in wild harvest activities in their traditional territory; the fishery served as the basis of an extensive trading network for the community, in addition to being central for food and in culture and ceremony (LaRiviere & Crawford, 2014; *R. v. Jones*, 1993 CanLII 8684 (ON SC)]). As Chief Nadjiwon of the Chippewas of Nawash Unceded First Nation said:

> Fishing has always been front and center as [a] food fishery, as a means of trade and barter. I mean, this is going pre-contact carrying on to today. It's always been a big part of the community . . . most families have some attachment to the fishery in one shape, way or form.

Today, SON has among the largest Indigenous commercial fisheries in the Canadian Great Lakes. There are approximately 30 commercial fish harvesters in the two communities, including seven tugs and four punts. Lake whitefish and lake trout comprise the majority of the harvest, with yellow perch, pickerel, ling, suckers, carp, salmon, rainbow trout, and lake herring caught in lesser numbers.

Randall Kahgee, Former Chief of the Chippewas of Saugeen described the relationship of the Saugeen Ojibway to the fishery:

> I think the fishery has always been an important part of our people. I think it's an expression. It's very much a part of our culture. Very much a part of our way of life. And I think it speaks to our authority and jurisdiction in our territory . . . for myself, seeing that fishery sustained both from an ecological and economic perspective is very important because it's very much a part of who we are. And we need to preserve that and protect that and safeguard that so those future generations will see the importance of that because just being out on that water, or walking those creeks is an exercise of not only our rights but our authority and our jurisdiction in our territory.

This sentiment was echoed by Paul Jones, community member and former member of the SON Joint Council: "It's more than just a livelihood, it's about you're tied to the land, you're tied to the water, you're tied to the animals, the plant system. It's, you're tied to who you are. So it's not just about fishing".

In the mid-1800s the Saugeen Ojibway signed a set of treaties with the British Crown (most notably treaty No. 72, known as the "Saugeen Surrenders", and treaty No. 451/2). While land surrenders were made, the Saugeen Ojibway did not surrender the possession or use of their fishing grounds, as recognized in the Royal Declaration of 1847 (Blair, 1997). Nonetheless, for generations, British colonial and later Canadian government policy served to restrict SON's fishing activity and management over their traditional territory (Koenig, 2005; Blair, 1997). Over subsequent decades, commercial and subsistence fishing became part of a mixed economy also involving farming and off-reserve employment; however, within the fishery, substantial abuses of power occurred as the settler government refused to recognize SON's ownership of the fishing resources (Koenig, 2005).

In 1993, in the context of escalating tensions between Indigenous and settler fishing communities, an Ontario Supreme Court decision was delivered in the Jones and Nadjiwon (Justice Fairgrieve) case, stemming from criminal charges by the Province of Ontario against fish harvesters from the Chippewas of Nawash. The decision reaffirmed SON's constitutionally protected right to fish for sustenance and trade within their traditional waters (Blair, 1997), and ruled that a ban on sales of fish imposed by the Province of Ontario on SON was unconstitutional [*R. v. Jones*, 1993 CanLII 8684 (ON SC)]. Randall Kahgee spoke about this reaffirmation of the right for SON:

> We should look at that case as a turning point where our people and our Nations now take full ownership of those rights, full jurisdiction and authority over those rights, and take our role as stewards seriously in that regard to ensure that those rights are sustainable for future generations. Because up until that point, it's not a question of whether we took those seriously, we were excluded from that conversation.

Emerging from this decision, SON entered into negotiations with the Ontario and federal governments concerning the management of commercial fisheries within their traditional waters. OMNRF also engaged in a process of "buying-back" quotas from non-native commercial harvesters in the traditional waters to make room for the SON fishermen. In 2000, SON and the province signed the first Fishing Agreement (the Agreement) covering key aspects of co-management including the full recognition of SON's Aboriginal and Treaty rights, territorial limits for commercial fishing, monitoring and compliance, information to be exchanged, and total allowable catches for commercial harvests (Akiwenzie & Roote, 2004). From OMNRF's perspective, the Agreement set out a cooperative framework to help ensure that both SON's commercial fishery and use of SON's traditional waters by recreational anglers (the other large resource user group on Lake Huron) continue to safely coexist, while maintaining a sustainable fishery (OMNRF, 2013). From SON's viewpoint there were still many issues that had not been resolved, but the Agreement was an important step towards ensuring that the Saugeen Ojibway could exercise their rights and their culture in their own territory, safely and without discrimination from the settler government or settler society.

In the most recent agreement signed in 2013 there was a substantial shift in focus from SON to begin reconciliation with the settler government, as explained by Randall Kahgee:

> The reconciliation work that should have happened after that case . . . I think it started and it's still continuing, and those things take time. Perhaps not as quickly as some may want, or our people may feel it needs to take, but it does take time. And the Crown has a significant role to play in that.

The 2013 agreement was based upon four pillars of critical importance to the SON. First, sustainability of the fishery in terms of protecting native fish species. Second, reopening access to Owen Sound and Colpoy's Bay for commercial fishing. In the early Agreements the Bays were off limits as a concession to the recreational fishery interest; however, this is difficult for SON's small punt fishery which needs the safety of the protected waters of the Bays. Third, addressing the impacts of fish stocking. The OMNRF has a long history of supporting the stocking of non-native fish species to satisfy the recreational fishing interest. SON believes that the stocking activities are causing harm to the native fish and thus SON's way of life, including their relationship to those original species. And fourth, revitalizing the economy of SON's commercial fishery and holding the Crown accountable for economic injustices. The Agreement does not cover fishing for food, ceremonial, or social purposes. The 2013 Agreement has been renewed several times, with the current Agreement in place until 2023.

SON is clear that entering into the Agreement does not erode their inherent Aboriginal and Treaty rights. Paul Jones described the Agreement as a mechanism for further exercising their rights to fish: "It's our right [to fish], and we're protecting our right so our children and our grandchildren down the road, they can still fish. It will be there". Randall Kahgee similarly explained:

> To be clear, we don't need the Agreement to fish. But I think it's important to have that because the Crown has to own up to that [injustice]. We can't let the Crown off the hook for allowing those injustice[s] to continue for over 150 plus years. Because if we do that we basically admonish them of their fiduciary obligation.

However, despite the Agreement and the importance of the fishery to their community, SON's participation in other levels of fisheries governance has not been recognized by the settler state. For example, SON has not been invited to sign A Joint Strategic Plan for Management of Great Lakes Fisheries, the multi-jurisdictional governance arrangement concerned with basin-wide fisheries management facilitated by the Great Lakes Fishery

Commission (GLFC). The Province of Ontario continues to assert they have the jurisdiction to represent all First Nations at the Commission. SON has spoken out against what they see as deliberate attempts to exclude them from this Commission (Akiwenzie & Roote, 2004, p. 24). In Chief Nadjiwon's words, "til' my dying breath I'm going to fight that whatever's in those waters we have an inherent right to harvest".

Discussion

BFN and SON's approaches to fisheries governance provide insights beyond only interactions among the First Nations and the state. First, these case studies question *who* holds power, as they raise concern about the legitimacy of the Canadian state and the Crown's assertion of sovereignty. Further, the expressed governance goals are about more than regulating fish as a resource. For both BFN and SON, governance is described as a process of building and maintaining relationships with the land and water along with the community's past, present, and future culture and identity. Ecological systems play an active role in these governance relationships. This approach to Anishinaabe governance expressed by both First Nations is different than the ideals of the settler state expressed through provincial policy and regulation. For both case studies, the interaction between the communities and their fisheries is a self-determined, context-based, and integrative form of governance. Both BFN and SON are committed to fishing in a way that is consistent with their cultural practices and considers the long-term sustainability of the ecosystems and their communities.

The case studies can also be considered in relation to the governance engagement continuum described in Chapter 1. This continuum is a useful heuristic tool for representing the ways that civil society and social movements engage in food systems governance *vis-à-vis* the state; this spans from multi-stakeholder arrangements (where the state retains ultimate control), to co-governance arrangements (involving some degree of power sharing between the state and other actors), to self-governance where Indigenous people, civil society groups, and social movement actors are the core participants exercising power in the governance of their own food systems. While Indigenous communities are key actors within food movements, they cannot simply be considered as a civil society or social movement actor but must be recognized as Nations with their own forms of government and unique aspirations. Nevertheless, the governance continuum can help in understanding these relationships with the settler state.

BFN and SON exhibit similarities and differences in their governance arrangements. In terms of similarities, both are committed to self-determination and asserting their inherent rights to fisheries. In relation to the governance engagement continuum, both BFN and SON exhibit tendencies towards self-governance. More specifically, both case studies show aspects of polycentrism as they seek to assert recognition of their existing self-governing processes within larger political structures and a settler state regime. The deployment and rebuilding of knowledge and inter-generational cultural resources disrupted by colonialism are important to maintaining their positions on the continuum (and not being only a stakeholder). However, a key aspect of a functioning polycentric governance system is coordination among the decision-making centres involved (Carlisle & Gruby, 2017); here, a lack of true Nation-to-Nation relationships hinders a polycentric system in both cases. In addition, decision-making centres at other scales, such as bi-national forums like the Great Lakes Fisheries Commission, are clearly not coordinated with the governance processes of First Nations.

Another similarity between BFN and SON is an expression that their inherent rights entail responsibility to protect all living things. Both communities consider the relationships with their fisheries well beyond a "resourcist" approach, seen in the linkages between fish, cultural practices, and ceremonies, and the use of traditional ecological knowledge. Decision-making is based on a long-term view of these relationships.

However, there are also differences, and specifically their relationships with the settler state. In terms of the governance continuum, SON may be seen as exhibiting characteristics of co-governance and self-governance, and BFN primarily showing characteristics of a self-governing arrangement. The Agreement established by SON with the Province of Ontario and the federal government aligns with Emerson and Nabatchi's (2015) characterization of collaborative governance as the processes and structures that engage people across sectors to "carry out a public purpose that could not otherwise be accomplished" (p. 18). The Agreement sets out an intention of cooperation between SON, the Province of Ontario, and different resource users around the Lake. In a spirit of collaboration, SON initially made certain voluntarily compromises under this Agreement, such as restricting fishing from the mouths of Owen Sound and Colpoys Bays to accommodate recreational anglers and boaters. While SON maintains they have ultimate ownership of resources, they have chosen to recognize a co-governance relationship with the Province and federal government as a step towards achieving their goal of self-governance, and as a means of holding the Crown accountable for past and present injustices. BFN, alternatively, continues to fish without negotiating the terms of a fishing license or formal agreement with OMNRF, asserting their traditional laws and practices as primary. From their perspective, cooperation with the settler state is a form of legitimation they don't agree with, and they continue to face tensions with the province and federal governments.

While addressing these governance relationships is vital to the future of BFN and SON, they are also key to emerging food sovereignty movements in Canada. Specifically, consideration of Indigenous fisheries governance structures, such as those illustrated in this chapter, may complicate interpretations of sovereignty movements in two key ways. First, food sovereignty scholarship has focused primarily on agricultural producers and agricultural-related policy and politics. As we have argued elsewhere (Levkoe et al, 2017), despite fishing communities playing a central role in food sovereignty movements, there has been limited integration of these experiences into research. Beyond making a vital contribution to existing debates and organizing, including small-scale fisheries complicates food sovereignty discourses by raising critical questions about the governance of common-pool food resources. Fish do not abide by political boundaries and national borders in the same way as land-based food production. In the case of BFN and SON, governing the Great Lakes fisheries means negotiating not only with various levels of the Canadian state but also with other Indigenous communities with overlapping claims to traditional fisheries, US-based Band Councils, inter-tribal management bodies, and state-level jurisdiction, as well as bi-national institutions such as the Great Lakes Fishery Commission – further complicating governance relations.

Second, as argued by Kepkiewicz and Dale (2018), the food sovereignty movement in Canada has primarily focused on settler aspirations, with the priorities of a settler agricultural regime dominating discussion. Considerations of forms of property ownership that are not private or other forms of agricultural and resource governance have received less attention; thus, there are important links that need to be built between Indigenous and settler communities within the movement. The case studies of BFN and SON present examples of fisheries governance that further complicate notions of food sovereignty in Canada and considerations that scholars and activists must wrestle with in respect to private property, political jurisdiction, and notions surrounding the control of nature resources. Furthermore, the two case studies demonstrate the complexity of what Indigenous rights over resources (e.g., food systems) means in practice. There is not a singular structure of what governance should look like for Indigenous communities; these cases highlight the diversity of arrangements that could be in place based on different traditions, needs, and desires. This further accentuates the demand for Nation-and-Nation governance relationships that must be considered with each Indigenous community. This is also an important message for settlers seeking to engage with broader visions of food sovereignty in Canada.

Conclusions

This chapter has presented two case studies of fisheries governance involving Indigenous and settler engagements, focusing on the Great Lakes region. Together, the two case studies demonstrate the complexity of rights, resources, and perspectives that comprise governance arrangements. Both communities, BFN and SON, are committed to self-determination and asserting their inherent rights to fisheries. They have also made different decisions that make sense for their own communities in relation to fisheries and governance. For settlers who may work with Indigenous people or read this chapter, we have attempted, not to judge these decisions, but rather to understand them in an effort to build meaningful Nation-to-Nation relationships.

These case studies bring attention to the need to go beyond a "resourcist" view on fisheries that continues to characterize state management regimes, and which generally seeks to manage people and the natural environment for aims of efficiency and profit (Olson et al, 2014; Berkes, 2010). In the two Indigenous communities in this chapter, fisheries themselves are expressed as a form of governance between people, culture, and nature. Thus, governance is not simply a structure to be imposed to regulate people and their relationships with the natural environment. Learning from the governance approaches embedded in Indigenous fisheries may be helpful for other communities and jurisdictions seeking to move beyond narrow technocratic management philosophies.

Lastly, these case studies also shed light on what rights over food systems from a food sovereignty perspective might look like in practice. This chapter complicates consideration of food sovereignty by bringing to the forefront fisheries as food systems – an element of food sovereignty that has been generally overlooked in the movement due to a focus on agricultural food production. The case studies of BFN and SON also illustrate questions that settlers, food sovereignty scholars, and activists must wrestle with in respect to private property, political jurisdiction, and notions surrounding the control of natural resources, particularly if the settler food sovereignty movement wants to challenge and not reinforce a history of Indigenous dispossession from land and water.

Acknowledgments

We sincerely thank all the participants who shared their time to contribute to sharing their knowledge and experiences; we especially appreciate the support of the Batchewana First Nation Chief and Council, and the Saugeen Ojibway Nation Joint Fisheries Committee and Joint Council. We also greatly acknowledge conceptual input from Connie Nelson, Gordon Hickey, and the Food: Locally Embedded Globally Engaged (FLEdGE) research team and financial support from the Social Science and Humanities Research Council of Canada.

Notes

1 In Canada, First Nation is the term used to identify Indigenous people (i.e. the first inhabitants of Canada) who are ethnically neither Métis nor Inuit; Métis denotes a collective of cultures and ethnic identities resulting from unions between Indigenous and European people in what is now Canada. However, First Nation does not have a legal definition. Here, the term "Aboriginal" is the legal term used to refer to Indigenous people (First Nations, Inuit, and Métis) as defined in Section 35 of the Canadian Constitution. In non-legal contexts, the term Indigenous is becoming increasingly preferred (Department of Justice, 2015; First Nations Studies Program, 2009).

2 While often used together, the terms self-determination and self-governance are distinct. Most simply, the right to self-governance may be understood as embedded in the broader right to self-determination. This relationship is clarified in Articles 3 and 4 of the United Nations Declaration on the Rights of Indigenous Peoples

(United Nations, 2008) which states: "Indigenous peoples have the right to self-determination. By virtue of that right they freely determine their political status and freely pursue their economic, social and cultural development. Indigenous peoples, in exercising their right to self-determination, have the right to autonomy or self-government in matters relating to their internal and local affairs, as well as ways and means for financing their autonomous functions".

3 In Canada, a reserve refers to a parcel of land set aside under the Indian Act and treaty agreements for the use of an "Indian Band" (a governing unit of Indians according to the Indian Act). Reserve borders were imposed by the settler state on First Nations, serving to undermine Indigenous peoples' relationships with their traditional territories (First Nations Studies Program, 2009).

4 Inherent rights are those believed to be granted to Indigenous people by the Creator and not granted by any particular Treaty, agreement, or law (Sanderson, 2017). These rights are related to but distinct from the "existing" Aboriginal rights reaffirmed by the Constitution Act of Canada (1982), Section 35(1).

5 As Canada is a constitutional monarchy, the Queen serves as head of state as the embodiment of the Crown.

6 In the case of inland waterways, the Federal Fisheries Act has devolved authority for fisheries management and decision-making to the provinces and territories.

References

Assembly of First Nations (n.d.) *Impacts of pollution on Great Lakes fisheries: Discussion paper*. Available at: www.afn.ca/uploads/files/env/great_lakes_considerations_paper.pdf

Akiwenzie, R. & Roote, R. (2004) *Jumping to conclusions: GLFC, conflict of interest and the problem of stocking exotic salmon in the Great Lakes*. Discussion Paper. IAGLR (International Association for Great Lakes Research) 2004 Conference, 24–28 May 2004, Waterloo, Ontario, University of Waterloo.

Berkes, F. (2010) Shifting perspectives on resource management: resilience and the reconceptualization of "natural resources" and "management". *MAST*, 9(1), 13–40.

Blair, P. (1997) Solemn promises and solum rights: the Saugeen Ojibway fishing grounds and R. V. Jones and Nadjiwon. *Ottawa Law Review*, 28, 125–143.

Borrows, J. (2010) *Drawing out law: a spirit's guide*. Toronto, University of Toronto Press.

Boudreau, S. A. & Fanning, L. M. (2016) *Fisheries management and decision making in Canada's inland waterways of Ontario*. Marine Affairs Program Technical Report #13. Available at: www.dal.ca/faculty/science/marine-affairs-program/research/research-news/map-technical-seriesreports.html

Brendan, T., Brown, R., Ebener, M., Reid, K., & Newcomb, T. (2012) Great lakes commercial fisheries: historical overview and prognoses for the future. In: Taylor, W., Lynch, A., & Leonard, N. (eds) *Great lakes fisheries policy and management: a binational perspective*. East Lansing, Michigan State University Press, pp. 339–398.

Carlisle, K. & Gruby, R. (2017) Polycentric systems of governance: a theoretical model for the commons. *Policy Studies Journal*. Available at: https://doi.org/10.1111/psj.12212

Christie, G. (2007) *Aboriginal nationhood and the inherent right to self-government*. Vancouver, National Centre for First Nations Governance.

Coulthard, G. (2014) *Red skins, white masks: rejecting the colonial politics of recognition*. Minneapolis, University of Minnesota Press.

Department of Justice (2015) *Logistics First Nation(s) – Aboriginal*. Available at: http://canada.justice.gc.ca/eng/rp-pr/csj-sjc/legis-redact/legistics/p1p12.html

Desmarais, A. A., Claeys, P., & Trauger, A. (eds) (2017) *Public policies for food sovereignty: social movements and the state*. New York, Routledge.

Emerson, K. & Nabatchi, T. (2015) *Collaborative governance regimes*. Washington, Georgetown University Press.

Environmental Commission of Ontario (2010/11) *Engaging solutions: annual report*. Available at: http://docs.assets.eco.on.ca/reports/environmental-protection/2010-2011/2010-11-AR.pdf.

First Nations Studies Program (2009) *Terminology*. Available at: https://indigenousfoundations.arts.ubc.ca/terminology/

Gaden, M., Goddard, C., & Read, J. (2012) Multi-jurisdictional management of the shared Great Lakes fishery: transcending conflict and diffuse political authority. In: Taylor, W., Lynch, A., & Leonard, N. (eds) *Great Lakes fisheries policy and management*. East Lansing, Michigan State University Press, pp. 305–338.

Government of Ontario (2015) *Ontario's provincial fish strategy: fish for the future*. Available at: www.ontario.ca/document/strategic-plan-ontario-fisheries

Hudson, J. & Ziegler, S. (2014) Environment, culture and the Great Lakes fisheries. *Geographical Review*, 104(4), 391–413.

Kepkiewicz, L. & Dale, B. (2018) Keeping "our" land: property, agriculture and tensions between Indigenous and settler visions of food sovereignty in Canada. *Journal of Peasant Studies*. Available at: doi.org/10.1080/03066150.2018.1439929

King, S. (2014) *Fishing in contested waters: place and community in Burnt Church/Esgenoopetitj*. Toronto, University of Toronto Press.

Koenig, E. (2005) *Cultures and ecologies: a native fishing conflict on the Saugeen-Bruce Peninsula*. Toronto, University of Toronto Press.

Kuhnlein, H. V. & Humphries, M. M. (2017) *Traditional animal foods of Indigenous peoples of Northern North America*. Montreal, Centre for Indigenous Peoples' Nutrition and Environment.

LaRiviere, C. & Crawford, S. (2014) Indigenous principles of wild harvest and management: an Ojibway community as a case study. *Human Ecology*, 41, 947–960.

LeBlanc, J. & Burnett, K. (2017) What happened to Indigenous food sovereignty in northern Ontario? Imposed political, economic, social-ecological and cultural systems change. In: Robidoux, M. & Mason, C. (eds) *A land not forgotten: indigenous food security and land-based practices in northern Ontario*. Winnipeg, University of Manitoba Press, pp. 16–33.

Levkoe, C. Z., Lowitt, K., & Nelson, C. (2017) "Fish as food": exploring a food sovereignty approach to small-scale fisheries. *Marine Policy*, 85, 65–70.

Lowitt, K., Johnston-Weiser, D., Lauzon, R., & Hickey, G. M. (2018) On food security and access to fish in the Saugeen Ojibway Nation, Lake Huron, Canada. *Journal of Great Lakes Research*, 44(1), 174–183.

Lytwyn, V. (1990) Ojibwa and Ottawa fisheries around Manitoulin Island: historical and geographical perspectives on Aboriginal and treaty fishing rights. *Native Studies Review*, 6(1), 1–30.

MacCrimmon, D. (2002) Sustainable fisheries management in the great lakes: scientific and operational challenges. *Lakes & Reservoirs: Research and Management*, 7, 241–254.

Martin, D. & Amos, M. (2016) What constitutes good food? Towards a critical Indigenous perspective of food and health. In: Koc, M., Sumner, J., & Winson, A.

(eds) *Critical perspectives in food studies*. Second Edition. Toronto, Oxford Press, pp. 205–220.

Morrison, D. (2011) Indigenous food sovereignty: a model for social learning. In: Wittman, H., Desmarais, A., & Wiebe, N. (eds) *Food sovereignty in Canada: creating just and sustainable food systems*. Halifaz, Fernwood Publishing, pp. 97–103.

Natcher, D. (2001) Co-management: an Aboriginal response to frontier development. *The Northern Review*, 23, 146–163.

Nyéléni Declaration (2007) *Declaration of Nyéléni. Sélingué, Mali*. Available at: https://nyeleni.org/spip.php?article290.

Olson, J., Clay, P. M., & Pinto da Silva, P. (2014) Putting the seafood in sustainable food systems. *Marine Policy*, 43, 104–111. Available at: doi:10.1016/j.marpol.2013.05.001

OMNRF (Ontario Ministry of Natural Resources and Forestry) (2013) *Backgrounder: fisheries agreement with Saugeen Ojibway Nation*. Available at: www.owensound.ca/en/resourcesGeneral/Documents/Fisheries-Agreement-with-SON---Backgrounder.pdf

OMNRF (2015) *Ontario's provincial fish strategy: fish for the future*. Available at: http://apps.mnr.gov.on.ca/public/files/er/ontarios-provincial-fish-strategy.pdf

Patel, R. (2009) Food sovereignty. *Journal of Peasant Studies*, 36(3), 663–706.

People's Food Policy (2011) *Resetting the table: a people's food policy for Canada*. Available at: https://foodsecurecanada.org/people-food-policy

Power, E. M. (2008) Conceptualizing food security for Aboriginal people in Canada. *Canadian Journal of Public Health/Revue Canadienne de Sante'e Publique*, 99(2), 95–97.

R. v. Jones, 1993 CanLII 8684 (ON SC). Available at: http://canlii.ca/t/g128h

R. v. Agawa, 1988 CanLII 148 (ON CA). Available at: www.canlii.org/en/on/onca/doc/1988/1988canlii148/1988canlii148.html?autocompleteStr=65%20Or%202d%20505&autocompletePos=1

Sanderson, S. (2017) *Constitution Act 1982: implementation of sections 35(1) and 35 (2)*. FSIN Office of the Senate. Available at: https://sencanada.ca/content/sen/committee/421/APPA/Briefs/ConstitutionAct_2017-09-19_e.pdf

Simpson, L. (2011) *Dancing on our turtle's back: stories of Nishnaabeg re-Creation, resurgence, and a new emergence*. Winnipeg, Arbeiter Ring Publishing.

Syrette, C., Darlene, S., & Corbiere, J. (1977) *Batchewana Band history, 1850–1977*. Algoma University archives. Available at: http://archives.algomau.ca/main/sites/default/files/2010-046_002_036.pdf

Tobias, J. & Richmond, C. (2014) "That land means everything to us as Anishinaabe . . .": environmental dispossession and resilience on the North Shore of Lake Superior. *Health & Place*, 29, 26–33.

United Nations (2008) *United Nations declaration on the rights of Indigenous people*. Available at: www.un.org/esa/socdev/unpfii/documents/DRIPS_en.pdf

8 Collaborative governance

The case of local food action planning

Jill K. Clark

Introduction

Trends in governance, as outlined in the introduction to this volume, emphasize the growing roles of the private sector and civil society in decision-making structures, with the normative assumption that greater participation is beneficial (Torfing, 2010; Minnery, 2007; Kooiman, 2003; Jessop, 1998). The food movement, as described in Chapter 1, is right in line with this thinking. Terms such as "food democracy" and "food citizenship" are being used in public discourse, calling attention to the role of citizens and civil society organizations (CSO)[1] in shaping the food system. Yet we know little about the mechanisms that empower, or disempower, community members and CSOs within networked decision-making structures around food systems (Moragues-Faus, 2017). For example, are civil society actors merely assigned stakeholder roles in governance structures, or are they collaborators in public problem-solving?

As such, the purpose of this chapter is to trace and analyze a case of local food planning within a unique governance structure in a United States (US) city – Columbus, Ohio. The governance structure, which included a city health department, a county economic development and planning department, and a local food movement CSO, was embedded in a local food network. The members of the co-governance structure were organized as collaborative planning partners that shared power in a local government planning process. The questions I explore in this case are the following: What led to a food movement CSO becoming a collaborative planning partner in a local government planning process? What does governance look like in practice, including structuring decision-making and influencing others? What is the resultant governance structure, with particular attention paid to the CSOs, post-planning? To answer these questions, I use document analysis (e.g., reports, minutes, web content), interviews with the planning team members,[2] and my own experience participating in the process.

Collaborative governance to collaborative governance regimes

Federal policy historically has played the primary policy role in shaping the dominant food system in the US. As such, local governments are exploring their role in creating place-based solutions to problems resulting from that system (Goddeeris, 2013; Scherb et al, 2012). Within new policy arenas, new forms of governance emerge, often including partnerships between the private, nonprofit and for-profit sectors along with the public sector and civil society groups (Mendes, 2008). Some research has suggested that networked governance may have less to do with downloading of responsibility to other groups (i.e., devolution), and more to do with non-governmental entities making or taking policy-making space, which in part reflects the growing expertise of civil society in this area (Raja et al, 2016; Renting et al, 2012). As such, civil society organizations, both formal and informal, play a significant role in local food system governance, particularly when these groups have territorial or placed-based interests (Ansell, 2003; Renting et al, 2012).

Governance of wicked problems gives rise to horizontal networks that build the capacity of local government to solve problems (O'Toole, 1997). These networks are distinct from hierarchical structures, where some organizations are formally subordinate to others. What "glues" networks together are common interests and values. Networks further the development of sharing understanding and meaning and provide venues for reconciling frames, or views of social reality (Sabel, 1993). The means of cohesion in networks is trust (Seppänen et al, 2007; O'Toole, 1997). This has been demonstrated many times in networks for policy development (Jenkins-Smith et al, 2014). Network-level trust rests on the trust between people and organizations that make the structure. Interactions that increase trust increase social capital, building local problem-solving capacity and the infrastructure for collective action (Musso & Weare, 2015).

Both the horizontal structure of networks and the mutual dependency and norms of reciprocity in networks mean that power is used to do work *with* others, versus used to wield *power over* others, and collaboration is a result of power sharing (Agranoff & McGuire, 2001). Therefore, while power is often not equal in the development of a collaborative, reciprocity and perceived fairness in power sharing are important. Networks that have a history of cooperation and some trust are more successful in implementing policies and programs (O'Toole, 1997). As Musso and Weare (2015) suggest, "these benefits may be self-perpetuating through a virtuous cycle in which social capital improves governance, which, in turn, promotes further civic engagement that enhances stocks of social capital" (p. 151).

Network conditions, along with other political and economic factors such as political dynamics and power relationships, history of conflict, and public service or resource conditions, can create enabling or disabling contexts for collaborative governance arrangements. As stated in Chapter 1, collaborative governance, as defined by Emerson and Nabatchi (2015), focuses on public policy decision-making and management and is made up of the processes and structures that engage people across sectors to "carry out a public purpose that could not otherwise be accomplished"(p 18). Members must be willing to conduct their work iteratively, rather than linearly, and have a commitment to this type of process (Ansell & Gash, 2008). Collaborative group decision-making usually moves towards consensus, and the responsibility for outcomes is shared (Ansell & Gash, 2008).

Collaborative governance arrangements can be the precursor to longer-term institutional arrangements or regimes. Collaborative governance regimes are forms of sustained cooperation between state and non-state actors and include "sets of implicit and explicit principles, rules, norms, and decision-making procedures around which actors' expectations converge in a given area" (Emerson et al, 2012, p. 6).[3] The drivers of these arrangements include: 1) the need to decrease uncertainty around a policy problem and the resources to address the problem now and into the future by increasing organizational stability; 2) key actors recognizing organizational interdependence in problem-solving; 3) consequential incentives (positive or negative) for potential collaborators; and 4) initial leadership that recognizes the first three drivers and uses their influence to create and populate a space that enables collaboration (Emerson & Nabatchi, 2015).

Collaborative governance arrangements, at their best, are opportunities to directly integrate non-state actors, thereby strengthening democratic process and problem-solving capacity.[4] The parties involved in collaboration may represent others or join as citizens or other stakeholders, likely depending on how these regimes form (Emerson & Nabatchi, 2015). For example, those that form by external authority (e.g., local government) are likely to have prescribed membership, whereas those that are grassroots-based will likely have voluntary participation, perhaps with recruitment. Those that are independently convened, such as by a third party that crosses sectoral boundaries, will likely form in a way that is attractive to diverse stakeholders to encourage participation. Once established, arrangements are dynamic and may change over time, but the initial form and conditions will continue to influence that evolution.

What is the role of food movement CSOs in collaborative governance regimes? Are such actors considered stakeholders or representatives of interests? On the one hand, social movement actors are often considered disruptive and adversarial, making change from the outside; on the other hand, contemporary forms of social movements are based on grassroots organizing and champion direct participatory democracy (Ansell, 2003). Grassroots organizations that are community-based are likely embedded in territorially based networks. Ansell (2003) finds that relational embeddedness in territorial networks that include multiple sectors can change attitudes towards collaboration versus an isolated homogenous network, which may be a better explanation for attitudes towards collaboration.

CSOs in the food movement, as described in Chapter 1, have different frames, work at different scales, and, therefore, have different approaches to change. For example, community food security is, by definition, concerned about embedding the food system in place, in a territorial community (Allen, 2004). Yet with community food security, a third tactic for change emerges, which is working to build alternatives outside the dominant system (Clark et al, 2015). A social movement organization must see its agenda accomplished through this governance process.

Laying the groundwork for collaboration

Franklin County, Ohio, is home to the capital city, Columbus. The population of the county is a little over 1.2 million (United States Census Bureau, 2015). On many metrics, the Columbus metropolitan area is doing well, with a robust, recession-proof economy, but at the same time, it is ranked as the second most economically segregated metropolitan area in the country (Williams, 2017; Price, 2015). This is coupled with food system-related problems, such as high rates of food insecurity and chronic disease, low fruit and vegetable intake, lack of access to healthy foods, lack of local markets for agricultural products, and unmanaged food waste.

These issues have been on the radar of local government for some time. One of the first government actions, in late 2009, was a regional food assessment and plan orchestrated by the Mid-Ohio Regional Planning Commission (MORPC). MORPC, which covers a 12-county region, helps local governmental members collaborate across borders on issues such as transportation and sustainability. The purpose of the plan was to assess the current state of the food system as a resource for decision-making. MORPC convened a working group composed of public sector staff (e.g., public health professionals, economic developers), non-profit social services, and food distributors and retailers. Other actors joined via task forces, including farm groups, gardening groups, university cooperative extension, food-related businesses, and other local government officials. No particular actors were identified to implement the plan.

Instead, the plan included a recommendation to create local food policy councils in each of the member counties. As such, MORPC fostered the development of the Franklin County Local Food Council (FCLFC). In September 2011, selected farmers, representatives of local non-profits, local agency staff (e.g., the city health department), local food business representatives (e.g., distribution, retailing), and educators agreed to create the FCLFC. The FCLFC is an independent, grassroots civil society group whose mission is "to expand, strengthen, and maintain a resilient and local food system in Franklin County and the surrounding area". The strategies used by the group have a lot to do with building relationships across the food system in a common effort to improve the community. This is done by hosting community forums to co-learn new topics, share information, and, eventually, weigh in on the local food planning process.

During this time, Columbus City Council and the Franklin County Board of Commissioners also were supporting local food initiatives. For example, both city and county governments independently supported community gardens, funded a pilot of a purchasing-incentives program for low-income residents at local farmers markets, helped launch a healthy corner store program, and provided regular funding for emergency food services. However, efforts between the city and county were not necessarily coordinated, and neither were those of the community groups receiving government dollars to change the local food system. As the FCLFC and community groups became more active, more requests for funding and other resources were made to elected officials. Recognizing the systemic aspects of these issues, and the need to coordinate efforts, the city and county launched a planning process in November of 2014. Therefore, the aim of the Local Food Action Plan (LFAP) was to provide a framework for common goals and actions to inform public policy and program development, facilitate community collaborations, and direct local funding (City of Columbus & Franklin County, 2016). A unique partnership was developed to ensure "economic development, health, equity, cultural competency and social justice were intentionally considered in all project phases". Typical food plans include city planners, other governance department staff, and private, for-profit consultants. While typically engaged in the planning process, CSOs are not usually part of the planning teams that develop the process itself.

Establishing the planning team and planning process

The two legislative sponsors that initiated the planning process were a county commissioner and a city council member. The department leading the county's effort was the Franklin County Economic Development and Planning department, and the city lead was Columbus Public Health. The lead planner from the county had championed several policy initiatives in the county, such as enabling the keeping and processing of chicken, ducks, and rabbits in residential areas and a policy resolution solidifying and strengthening the county's commitment to supporting the local food system. Shortly after the launch of the planning team, a second county planner joined. The health department had extensive experience working with community members on identifying areas that lacked access to healthy food. The lead from the health department of was joined by two other staff members.

The third organization that was part of the planning team is the non-profit Local Matters, which was brought on at the request of the city legislative lead and hired through the health department. Local Matters is a food-movement CSO based on the concept of community food security, which is defined as "a situation in which all community residents obtain a safe, culturally acceptable, nutritionally adequate diet through a sustainable food system that maximizes community self-reliance and social justice" (Hamm & Bellows 2003, p. 37). Groups adhering to community food security as an approach to addressing problems in the food system have long been part of the alternative agrifood movement (Clark et al, 2015; Allen, 2004). Created in 2010, the Local Matters mission is "to create healthy communities through food education, access and advocacy". In its first annual report, it notes that it distinguishes itself by connecting the different parts of the food system by building trusting and long-term relationships with other community organizations that share its values and by listening to the wisdom of neighbours that live in the community. Further, this organization includes language regarding social equity in the food system in its long-term vision.

Local Matters had several years of experience working in the community, although it should be noted that the director of community outreach had been organizing around food system issues in the community since 1998 and was part of the MORPC regional planning process mentioned earlier. This long history of community organizing, relationship-building, and programming has made Local Matters a "go-to" organization when it comes to community food issues. The executive director and director of community outreach were on the LFAP project team, joined later by a hired consultant who was a member of the local food council.

The project team developed a working committee to help guide the process. The working committee was made up of 24 community members representing a variety of food system sectors (e.g., production, distribution, retailers) and other sectors, such as social services and education, although most members were already engaged in local food system work. Others would be involved in the process, as described in the next section. One thing to note about the project team and the working committee is the crossover in formal relationships. For example, several members of the project team and the working committee were active members of the grassroots local food policy coalition, the Franklin County Local Food Council.[5] In addition, the CSO had experience partnering with the health department, and staff of the health department were also part of the local food coalition. These are just a few examples of how the planning team was positioned within a broader network.

The local food action plan and plan process

The team developed the plan in 3 phases, which, after 24 months, produced a plan with 4 goals and 27 recommended actions (City of Columbus & Franklin County, 2016). Current needs and assets were documented in the first phase, and a vision was created. Neighbourhood meetings in six of Columbus's most impoverished neighbourhoods and one upper-income neighbourhood were held to identify neighbourhood-specific problems in access to healthy food and nutrition education. Four different online surveys aimed at different food system sectors, consumers, producers, processors/distributors, commercial and institutional buyers, and waste recovery were used to collect other primary data, resulting in approximately 700 responses. Finally, about a dozen key informants were interviewed to help with problem setting.

Secondary data collection for the first phase included compiling and analyzing published data, such as the population and agricultural censuses, to describe social and economic conditions (City of Columbus & Franklin County, 2016). In addition, team members compiled and analyzed local, regional, and state food action plans and public health reports from across the country. At the very end of the first phase, the working committee met to review draft documents and establish a vision. The resulting vision is that "a fair and sustainable food system that benefits our economy, our environment, and all people". The four goals to achieve the vision are to 1) enhance coordination and communication among existing food resources and agencies; 2) improve access to and education about healthy food, affordable food, and local food; 3) increase the role of food in economic development; and 4) prevent food-related waste.

The second phase identified potential actions to achieve the vision and goals. Actions identified in the scan of national plans and reports, the neighbourhood meetings, and previous community planning documents seeded this phase. The national scan and neighbourhood meetings resulted in hundreds of actions. Over the course of two meetings, these actions were reviewed by the working committee, narrowing them down to the 44 most promising actions that were impactful and could be implemented. The planning team conducted research on good practices for each one, combining some actions, which resulted in 35 actions. The working committee reviewed the draft at a final meeting, which was open for feedback from the public, and chose a final set of 27 actions to be included in the final plan.

The final phase of the project included developing a "Getting Started" table for each action, which included guidance on plan implementation that identified potential public and private resources available for implementation, starting steps, and measures for evaluation. Once complete, the working committee reviewed the full draft LFAP. After the go-ahead from the commissioner and city council member, a six-week public comment period opened.

The public comment period included both a general public online survey and 11 feedback sessions where the team travelled to community and neighbourhood meetings. During these meetings, actions were prioritized into those that would have the greatest impact and those that were "low-hanging fruit". Further, during this period the Franklin County Local Food Council spent its meetings reviewing the plan, dedicating each meeting to a general area of the plan. The council began to map how it saw the council contributing in a community governance structure to help guide implementation of the plan. Project team members, some of whom are members of the council, participated alongside council members.

Governance of the planning process

The governance of the planning process emerged from a long history of food-related projects in the community. This section describes the origins of the governance arrangement as well as the roles of, and relationships among, planning team members and other stakeholders. It also examines the dynamics of the decision-making processes within the arrangement.

Origin stories

When asked how the joint city–county–CSO planning process came to be, each interviewee shared a different explanation. All interviewees but one started their narratives at the point when the planning process was first being discussed in their organizations. The interviewee who did not start at that point instead began with a narrative of activity that occurred two decades prior to the start of the planning process. When the other interviewees were pushed to answer what precipitated the possibility of the plan, they also provided different explanations. However, all explanations were based on the same three themes. First, people and organizations in the community had been doing work in this area for quite some time. Second, this work was embedded in a set of networked relationships between government, CSOs, community members, and some private, for-profit actors. Some groupings in the broader community network were formalized coalitions, such as the regional food council, the local food council, and the healthy food access coalition, all of which included project team members and organizations. Team members also had personal relationships based on having done projects together. For example, one member of the CSO was contracted 13 years earlier to work with the city's health department as it began to explore the role of food in public health. Each team member could trace multiple first- and second-degree relationships through the network back to other team members. Third, this network was bound by a commitment to addressing needs in the food system, which gave rise to a discussion of the need to coordinate action. All of the stories ended with the legislative sponsors committing to the planning process. In essence, the informal governance of local food systems set the stage for formal governance. One interviewee stated:

> I think one of the reasons this was so successful is that the process, at least for me personally . . . started 20 years ago, so I don't know how you capture that but this kind of work, I think what a lot of people need to fully understand is the relationship building that is so key up to when we get into a more formal basis.

One big hairy mess of collaborative decision-making

In 2014, the legislative sponsors (county and city) each chose a department and a lead to begin the planning process. In addition, the city council member wanted the CSO to be contracted by the health department because of the CSO's strength in the community engagement process and its prominence in the local food movement. When asked what the organizational structure of the project team would look like on paper, one interviewee summed it up: "It would look like a big hairy mess [laughs]. It literally would just be a pool of people". No one party was designated to run the process; "It was never defined". The city and county legislative sponsors wanted all things to be equal. The staff had different reporting structures that did not interface, so they were on equal footing. The CSO was on contract with the city, but as a result of a previous relationship with decision-makers and staff at the department of health, the city treated the CSO as a collaborative partner (albeit one that was still responsible for reporting to the city), creating a flat organizational structure. As described earlier, the power structure does not have to be equal in the development of a collaborative, but reciprocity and perceived fairness in power sharing is what made this work as a flat structure (more on this later):

> The project team structure on paper would be probably like an amoeba . . . these collaborative type of projects between the city and the county, they're just unique. The city can't tell the county what to do and the county can't tell the city what to do. We all have to be there and participating in a collaborative spirit with the understanding that we're all trying to work towards the same end. We might have different reasons for getting there, but we're all trying to get to the same end place.

When it came to the work, team members shared responsibility for managing and facilitating project team meetings and producing the final products. The following reflects a shared sentiment among planning team members:

> Anything that was done, was done very intentionally through this project. If someone on the project team didn't agree with it, they brought it up and we hashed it out in project team meetings. To hear at least the three viewpoints that were in that room, to hear those out as far as the best way forward, and really ended up with a better product at the end of it because of that just deep thinking from those different perspectives that it wasn't just one entity or one viewpoint saying this is the best thing for us to do in the project to move it forward.

Eventually, project team members would take the lead on pieces of the work that matched their skill sets. For example, some members were more skilled at facilitation, others at tracking progress, and others at research. Each interviewee noted that the other two organizations brought topical expertise, such as public health, community food security, and economic development. Each organization in the project team had expertise in different areas of the process:

> If it was, "How does this phase of the planning process work?", then Franklin County had a lot of opinions about that, but if it was, "How do we sort through all of these varying claims and competing actions and all of that?" then Columbus Public Health wanted to wrestle with all those things and think about integration . . . Local Matters was constantly trying to provide the community place and say, "How will this work for the grass roots? How will this work for the non-profits? How will this work for marginalized and under-represented communities?"

With legislative sponsors wanting a finished project in an 18- to 24-month period, there was pressure to figure out how to get the plan done in that timeframe. As a result, the project team met weekly and convened numerous half-day work retreats. Even though no one on the project team knew every other person on the team prior to starting the work, it was able to begin collaborative work very quickly. Reasons for the quick start had to do with a level of trust as a result of relationship-building between team members prior to the team coming together.

Prior relationships and pressure from the legislators to make it work, together with frequent meetings, led to collaborative decision-making. Even the newest member to join the team, who considered himself to be the person with the least authority, felt he had as much input and influence as the other team members. As another team member put it: "It was very collaborative, very like loose, if you will, kind of like there weren't, there weren't severe or significant boundary lines between what one person did and what they didn't do so". The trust that developed allowed team members to be vulnerable in front of one another, including being willing to learn from each other, to challenge others' ideas (but not the person), to throw out wild ideas, and to trust others' work.

According to all the interviewees, aside from relationships within the broader network, the glue that kept the team together was a deep commitment and belief that the food system is a public commons:

> Well everyone believed that food was important and it had, it had profound implications, on how we handle food and agriculture had implications on the communities that we were working for, that we were, whether by choice or because that was our job. So, the unifying thing was that we were there to help communities that we were already serving on a daily basis.

While there were disagreements, they were process-oriented or topical, not relational. The relationships and the "underlying commitment" of the team to food and the community meant that team members were willing to reconcile difficult matters. One major disagreement was about the topic of livable wages in the plan, which eventually came to a compromise.[6]

> I think it's because we truly had total respect for each other and we knew whatever the perspective was it wasn't coming with an agenda, it was coming with true care for the community and I think we just really respected each other intellectually and personally and so we could do that.

It should be noted, however, that this collaborative decision-making and lack of a hierarchy did come at a cost, namely time that had to be invested to navigate new governance terrain and negotiate even on the smallest of decisions as a result of differing approaches and perspectives.

Plans change

The collaborative decision-making and negotiation of different perspectives certainly applied to planning and implementing the public participation component of the planning process. It started with defining a problem setting, which, thereby, drove the planning process, including who should be a part of it.

> When we started talking to the city, it just became that the need within the community was so much more than just this economic development review of things and we really needed to identify current conditions throughout the entire system . . . That wasn't our vision going into it, but we just recognized that there were these issues within the food system that were systemic and county wide that we needed to address at a city/county level if we wanted to effect any change with them.

Each organization had plenty of experience working with different types of "publics", including the public-at-large, stakeholders and stakeholder groups, experts, and community members. Also, each organization had a different approach. For example, the county approach could be described as standard planning practice, both structured and linear, and likely to occur in government spaces with a focus on consulting. The CSO approach could be described as social justice-oriented, likely to occur in impacted communities, focus on listening, and involve programming or projects the communities requested.

With no official procedures in place for the team, a flat organizational structure, and a collaborative decision-making process, a way forward was negotiated. It took time to develop the "new language" for public participation. Each project team needed to learn how to translate its process norms to the other organizations so that the other organization saw the value of these norms. Talking about this translation, one interviewee said, "Hey, you need to make your spreadsheets so we understand them", and, "Hey, if we're going to do touchy feely stuff, it can't just be fuzzies for the sake of fuzzies. It's got to be with a goal in mind".

What resulted was a process unlike any that the three entities used independently. While the planning phases provided an overall structure that kept the team moving forward, there were many feedback loops within each phase. Through team deliberation, the teams placed an emphasis on regularly asking themselves if they were hearing from the public and incorporating their needs, which required adjusting approaches midstream while maintaining the overall planning structure. This meant adding new voices to the process and approaching new venues for public participation. It also meant changing agendas by using feedback from participants. For example, the working committee that operated during the entire planning process interacted regularly with the team, impacting the steps taken in the process.

> Standard planning processes is that you're kind of there to get feedback from them, almost like extract what you need from them. But this working committee was meant to be more of a, like a give and take or they, they're kind of almost an extension of the project team.

Deliberation also resulted in an emphasis on building the relational capacity of the community by arranging meetings so that working committee members interacted with members they did not know.

Post-plan governance

The planning process resulted in governmental support for continuing to build and strengthen local governance, both formal and informal, towards implementation. By expanding the usual suspects in the working committee and focusing on relationship building, the informal network grew. In addition to this growing social infrastructure, the planning process also resulted in a formal governance regime that marries government-led structures with grassroots community efforts.

The planning process resulted in an entire section of the plan aimed at "Enhanc[ing] coordination and communication among existing food resources and agencies". As part of that plan section, the city and county established, and have already staffed, a permanent joint Local Food Team. The Local Food Team facilitates an appointed, 12-person board (six from the county and six from the city) that is charged with implementing the plan. This private–public collaboration leverages public and private resources, reports to the community, and coordinates efforts. The board members, in part, included members from the working committee that requested to be on the board. Further, on the recommendation of the grassroots, independent local food council, the Local Food Team will provide administrative support to strengthen the council. The city and county envision the local food council as a pathway to public participation, ensuring a connection to the community, and providing volunteers for ad hoc committees. By building its capacity, the council will be in a better position to achieve desired

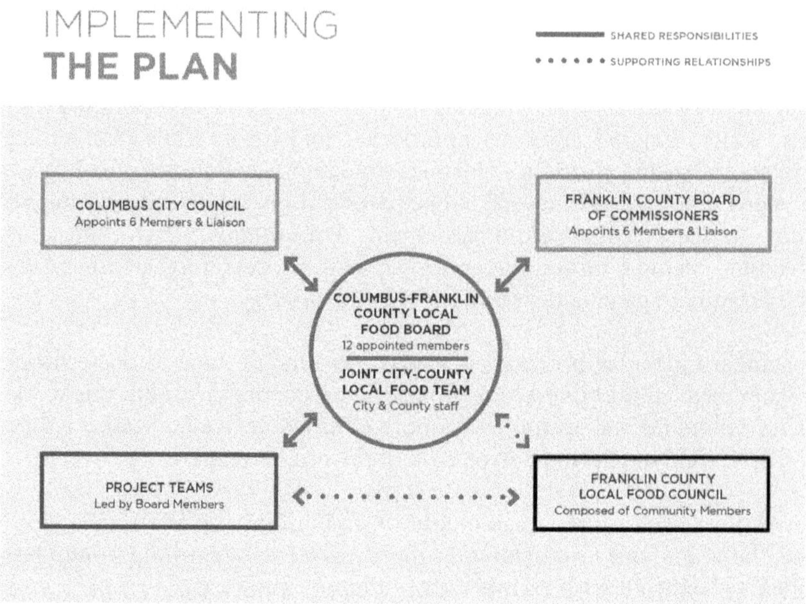

Figure 8.1 Local food action plan governance arrangement
Source: City of Columbus

social equity outcomes. Further, the council will identify stakeholders and interested parties for implementation. The plan is considered a "living document", as conditions are always changing, community work continues, and new opportunities arise. As such, the local food council has provided updates to the board regarding plan-related activities that are being conducted by community members.

Finally, the LFAP planning process increased the capacity of the planning team, which builds the capacity of local governance. One city team member has taken a city staff position on the Local Food Team. Two of the three members of the Local Food Team were part of the project team that developed the LFAP. They are embedded in local government departments that are supervised by two other members of the project team. All three members of the CSO are involved – one as a member of the Local Food Team advisory board, one as a member of the local food council, and one as a newly hired county staff person in the Local Food Team. This new hire said that through the planning process he learned the power of civil service, despite the bureaucracy (or because of it) in working for the good of the community. Every team member interviewed, unprompted, said that the LFAP planning process was the best professional experience they have had. They spoke of topical learning, including what it means to learn about food system economics and its relation to public health and to learn about the planning process. Each interviewee commented on how the perspectives that each of the other two organizations brought to the process were critical in the development of the innovative public participation process. Finally, all interviewees recognized room for improvement, whether on how to wrap the process back around to the individual neighbourhoods or how the use of a neutral facilitator would have made the team meetings more efficient.

Networks to regimes

Questions about how a food movement CSO becomes a collaborative planning partner in a local government planning process led into this case, including what the process looks like and what type of governance structure results (if at all). According to the national scan of food action plans, no CSO has been a collaborative designer of a local planning process with local governments. The case of the LFAP illustrates how the formal governance of the planning process was undergirded by a long-standing network of public, private, and civil society actors who had a commonly held belief that food is a vehicle for community betterment.[7] The role of commonly held beliefs is often discussed as part of the policy process (Jenkins-Smith et al, 2014). As was evident in the different origin stories that were based on common themes, the unique combination of project team organizations and actors emerged from informal networks of trust and community. The social infrastructure, or scaffolding of relationships, facilitated the local government planning process and the unique partnerships.

When you add the city–county partnership, with each entity backed by a legislative sponsor wanting to share equally, you end up with a flat organizational structure, despite the power differential with the food movement CSO. While initially the legislative (elected) leads used their coercive, instrumental power to create the space for the new governance arrangement, power was exercised *with and between* project team members embedded in networks laden with norms of trust and reciprocity. Perceived fairness in the process overcame differences in power between the food movement CSO and the two government organizations. For example, the process for decision-making within the project team (structural power), and the resultant agendas, were the result of a consensus-based collaborative project. Further, new norms and a “language” for what a planning process is (discursive power) were negotiated. One interviewee said, “there emerged a common language between [planning team members] of how to talk about the planning process, how to talk about what we meant when we talked about planning".

All the drivers were present for the emergence of a collaborative governance regime. Once public health, economic development, and community engagement interests deliberated on the nature of food system problems from their perspectives, it was evident to the team members that implementing a plan would require the whole of the network. As the purpose of the plan was to guide future action and investments, key stakeholders with beliefs in community betterment through the food system recognized the consequential incentives for collaborative participation. Finally, the local government legislative leads provided the legitimacy and leadership to create the regime structure. While ultimately the local government arguably had considerably more resources and decision-making power, the trust that non-governmental collaborators had in the process of decision-making and implementation of the plan compensated for that differential.

The resulting implementation governance structure positions the CSO and independent, grassroots food councils in a co-governance arrangement with the city and county solidly in the middle of the spectrum introduced in Chapter 1. The positioning, I argue, is the result of the CSO and the broader network (including government actors) making or opening space for themselves over a long period of time.

The implementation of the plan is facilitated by a joint city–county team that has a commitment to continual collaboration with a range of private actors (e.g., community members, non-profit organizations, for-profit firms) through its advisory board and to supporting an independent, grassroots food council. Because of this increased capacity, the food council will have four main responsibilities: 1) increase its membership and broader network to include the neighbourhood members that were part of the initial planning process and other stakeholders; 2) use its network to identify community members that should be part of the design of the implementation of each of the 27 plan actions; 3) keep the plan "alive" by updating the city and county on community action; and 4) hold the city and county accountable to the plan. Part of the rationale behind the "checks and balances" in this structure is a response to struggles in other communities. Formal governance structures can co-opt social movement actors or stifle protest voices (Newman et al, 2004). For example, the Portland Multnomah Food Policy Council, in part, was dissolved as a result of fully institutionalizing local food policy work within local government by hiring full-time staff positions, thus making the council's expertise less relevant (Coplen & Cuneo, 2016). At this point in time, the resulting governance structure of our food council, therefore, has the legitimacy of government, community groups, and businesses; the authority of local government; accountability to both the government and the community; resiliency to leadership change; and the strength of the growing social infrastructure.

We, of course, do not know the long-term effectiveness of this structure at this time. We do know that the city and county continue to support and provide resources for the efforts of the staff of the local food team. This is not a minor note, as the process described here was well supported and well funded by political leaders. Further, the CSO was paid to engage in the governance arrangement during the planning process. (The organization is not paid now, but the executive director is a member of the food board.) The food board described above serves on a voluntary basis. The food council continues to be a group of grassroots volunteers, but it has the benefit of eight hours a week of a local food team staff person's time.

Conclusion

The case of the local food action planning process illustrates the important role of informal networks built on trust and a common commitment to empower a food movement, civil society organization to be a partner in a co-governance arrangement. This suggests that the first step in a collaborative governance planning process is not planning, but focusing on building relationships, trust, and shared values. Just as important is the commitment of the two local governments to legitimize, enable, and structure the governance space. These governments recognized both the skills of their existing staff in the city health department and the county economic development and planning department and the expertise in the community. As such, they incentivized a collaborative arrangement within which co-learning could take place and power sharing was expected. It is important to note that while collaborators in the arrangement were committed to co-learning and power sharing in the decision-making process, this arrangement did come with costs, namely time and energy to devote to the process.

Being a collaborator in the planning process is not the same as being a partner in the implementation process. The values of civil society partners are embedded in the design of the process, including planning and how public decisions are made. In this case, those values included social justice and community empowerment lenses. The local food action planning process resulted in an expanded community network and an innovative governance structure for implementation that includes reciprocal relationships between local government, the private sector, and a grassroots, civil society coalition. Consequently, the capacity of the network in the city of Columbus and Franklin County to tackle wicked food system problems has been increased.

Acknowledgements

Thank you to the members of the LFAP planning team for allowing me to tell your stories. I appreciate your willingness to donate your time and your enthusiasm for this research. I thank Mary Beckie and Ana Moragues-Faus for reviewing an early draft, and Charles Z. Levkoe and Peter Andrée for their helpful direction for revisions of my final draft.

Notes

1 By civil society organization, I refer to the introductory chapter definition: "include a multiplicity of formal and informal associations, religious organizations and social groups – as distinct from governments and businesses – that work for what they understand to be the collective interests of society".
2 The planning team members read this chapter and provided additional insights and clarifications. There were no disagreements about the content of this chapter, but team members did provide a few corrections and elaboration on some points.
3 For a more exhaustive discussion of collaborative governance regimes, see Emerson and Nabatchi (2015) and Emerson et al (2012).
4 At worst, collaborative governance can be a barrier to the democratic process. For example, see Booher (2004).
5 The author was a working committee member and previously co-chaired the local food policy coalition with the non-profit director of community outreach.
6 A member of the working committee that represents farm workers wanted to include an action item in the plan to address livable wages. The CSO members of the planning team were in support of including this action item, but the local government planning team members argued that livable wages was outside of the scope of the plan. They compromised by including an action to "study and describe the impact of wages and policies on food system workers and the affordability of food".
7 While private, for-profit firms and actors have engaged in all the processes described in this case, these actors arguably have not played as significant a role in organizing, leading, or implementing these processes.

References

Agranoff, R. & McGuire, M. (2001) Big questions in public network management research. *Journal of Public Administration Research and Theory*, 11(3), 295–326.

Allen, P. (2004) *Together at the table: sustainability and sustenance in the American agrifood system*. University Park, Pennsylvania State University Press. Published in cooperation with the Rural Sociological Society.

Ansell, C. (2003) Community embeddedness and collaborative governance in the San Francisco Bay area. In: Diani, M. (ed.) *Social movements and networks: Relational approaches to collective action*. New York, Doug McAdam, pp. 123–144.

Ansell, C. & Gash, A. (2008) Collaborative governance in theory and practice. *Journal of Public Administration Research*, 18, 543–571.

Booher, D. E. (2004) Collaborative governance practices and democracy. *National Civic Review*, 93(4), 32–46.

City of Columbus & Franklin County (2016) *Local food action plan*. Available at: www.columbus.gov/publichealth/programs/Local-Food-Action-Plan/

Clark, J. K., Sharp, J. S., & Dugan, K. L. (2015) The agrifood system policy agenda and research domain. *Journal of Rural Studies*, 42, 112–122.

Coplen, A. K. & Cuneo, M. (2016) Dissolved: lessons learned from the Portland Multnomah food policy council. *Journal of Agriculture, Food Systems, and Community Development*, 5(2), 91–107.

Emerson, K. & Nabatchi, T. (2015) *Collaborative governance regimes*. Georgetown, Georgetown University Press.

Emerson, K., Nabatchi, T., & Balogh, S. (2012) An integrative framework for collaborative governance. *Journal of Public Administration Research and Theory*, 22(1), 1–29.

Goddeeris, L. (2013) *Local government support for food systems: themes and opportunities from national data.* Available at: http://foodsystems.msu.edu/uploads/files/local-govt-survey-brief.pdf

Hamm, M. W. & Bellows, A. C. (2003) Community food security and nutrition educators. *Journal of Nutrition Education and Behavior*, 35(1), 37–43, doi:10.1016/S1499-4046(06)60325-4

Jenkins-Smith, H. C., Nohrstedt, D., Weible, C. M., & Sabatier, P. A. (2014) The advocacy coalition framework: foundations, evolution, and ongoing research. In: Sabatier, P. A. & Weible, C. M. (eds) *Theories of the policy process.* Boulder, Westview Press, pp. 183–224.

Jessop, B. (1998) The rise of governance and the risks of failure: the case of economic development. *International Social Science Journal*, 50(155), 29–45.

Koc, M., MacRae, R., Desjardins, E., & Roberts, W. (2008) Getting civil about food: the interactions between civil society and the state to advance sustainable food systems in Canada. *Journal of Hunger and Environmental Nutrition*, 3(2/3), 122–144.

Kooiman, J. (2003) *Governing as governance.* London, Sage.

Mendes, W. (2008) Implementing social and environmental policies in cities: the case of food policy in Vancouver, Canada. *International Journal of Urban and Regional Research*, 32(4), 942–967.

Minnery, J. (2007) Stars and their supporting cast: state, market and community as actors in urban governance. *Urban Policy and Research*, 25(3), 325–345.

Moragues-Faus, A. (2017) Emancipatory or neoliberal food politics? Exploring the "politics of collectivity" of buying groups in the search for egalitarian food democracies. *Antipode*, 49(2), 455–476.

Musso, J. A. & Weare, C. (2015) From participatory reform to social capital: micro-motives and the macro-structure of civil society networks. *Public Administration Review*, 75(1), 150–164.

Newman, J., Barnes, M., Sullivan, H., & Knops, A. (2004) Public participation and collaborative governance. *Journal of Social Policy*, 33(2), 203–223, doi:10.1017/S0047279403007499

O'Toole Jr, L. J. (1997) Treating networks seriously: practical and research-based agendas in public administration. *Public Administration Review*, 57(1), 45–52.

Price, R. (2015) Neighborhood inequality particularly profound in Columbus area. *Columbus Dispatch*, 12 July. Available at: www.dispatch.com/content/stories/local/2015/07/12/neighborhood-inequality-particularly-profound-here.html

Raja, S., Picard, D., Baek, S., & Delgado, C. (2016) Rustbelt radicalism: a decade of food systems planning practice in Buffalo, New York (USA). *Journal of Agriculture, Food Systems, and Community Development*, 4(4), 173–189.

Renting, H., Schermer, M., & Rossi, A. (2012) Building food democracy: exploring civic food networks and newly emerging forms of food citizenship. *International Journal of Sociology of Agriculture & Food*, 19(3), 289–307.

Sabel, C. F. (1993) Studied trust: building new forms of cooperation in a volatile economy. *Human Relations*, 46(9), 1133–1170.

Scherb, A., Palmer, A., Frattaroli, S., & Pollack, K. (2012) Exploring food system policy: a survey of food policy councils in the United States. *Journal of Agriculture, Food Systems, and Community Development*, 2(4), 3–14.

Seppänen, R., Blomqvist, K., & Sundqvist, S. (2007) Measuring inter-organizational trust: a critical review of the empirical research in 1990–2003. *Industrial Marketing Management*, 36(2), 249–265.

Torfing, J. (2010) Local forms of governance in Denmark: the revenge of the supplement. *Journal of Power*, 3(3), 405–425.

United States Census Bureau (2015) *Population estimates*. Available at: www.census.gov/popest/data/cities/totals/2014/SUB-EST2014.html

Williams, M. (2017) Columbus is among few Midwest metro areas with robust economy. *Columbus Dispatch*, 29 May. Available at: www.dispatch.com/news/20170529/columbus-is-among-few-midwest-metro-areas-with-robust-economy

9 Hybrid governance as rural development

Market, state, and civil society in Correns, France

Chantal Wei-Ying Clément

Introduction

Rapid urbanization has become a defining feature of the 21st century. Rural populations are flocking to cities around the world in search of greater socio-economic opportunities, exacerbating urban sprawl and the growth of mega-cities. By 2050, the United Nations (UN) estimates that around 66 per cent of the global population will become urban (United Nations Habitat, 2016). Beyond the clear repercussions on ensuring food security, these trends raise a series of serious implications for food system sustainability, precipitating a plethora of proposals from academic and practitioners alike on how to address the demands of an urbanizing world.

Solutions that explore the possibility to transform and revitalize rural spaces are often left by the wayside in favour of prioritizing the values and objectives of urban spaces. For over half a century, industrial agricultural and food systems have developed to the detriment of rural areas. Alongside modernization and growth, many rural communities have experienced not only economic loss, but a loss of purpose and identity as well. In many ways, food system practice and research has become "too urban" (McMahon, 2014, p. 126).

Yet, in Europe as elsewhere, rural spaces continue to play an essential role in providing people with community, identity, and economic livelihood, while remaining critical sites for innovation. Rural communities – often by the very nature of their smaller population size and area – can more easily facilitate the creation of certain governance innovations that exist less often in larger, densely populated spaces. This chapter explores a hybrid of market and collaborative governance schemes developed in Correns, France's "first organic village", to revitalize the small rural community. In particular, I consider the role of farmers and civil society in the design and implementation of local governance strategies, and specifically, how a lack of distinction between these two sets of actors and the institutional sphere allowed for coherent and cohesive community development.

As raised in the introductory chapter to this volume, a lingering weakness within food governance research has been its inability to fully acknowledge the real-world implications of new governance arrangements and how they occur in practice (Paquet & Wilson, 2011). A more nuanced understanding of the new and innovative governance dynamics being practiced on the ground would allow for greater reflection on the hybridities within governance approaches often required to develop alternative food systems (e.g. bridging the divide between "local food" and global trading schemes, considering constructive applications of sustainable capitalism alongside more transformative goals). Further, focusing solely on private or community-based governance schemes runs the risk of missing key opportunities of state intervention and support, while only considering state-based initiatives ignores important market and social interactions that shape food systems.

The case of Correns is of notable interest as it relies on a hybrid governance approach, defined here as a form of collaborative governance that goes beyond the mere co-existence of, or competition between, different governance structures. In Correns, hybrid governance merges various types of arrangements, as well as both reformist and alternative orientations towards the dominant food system; on the one hand, Correns relies on value-adding market schemes through certifications including organic or *Appellation d'Origine Controlée*[1] (AOC) labelling, and on the other, focuses on collaborative state–community initiatives. Further, what began as Correns's decision to transition towards organic agricultural practices to improve economic livelihoods (i.e. the market approach) grew to encompass much broader socio-cultural and political goals (i.e. the collaborative state–community approach). Ultimately, Correns's search for economic viability paved the way for an integrated and democratic process for rural community development, allowing municipal officials and villagers alike to recognize the economic, social, and environmental value of their local food system, as well as its capacity to support vibrant rural livelihoods.

This chapter illustrates how hybridized governance not only proves a key enabler to redress current food system failures, but also provides a much-needed tool for rural revitalization adapted to a community's specific socio-cultural context. After an overview of current food systems' primary challenges and the local alternative food system arrangements seeking to redress them, this chapter highlights Correns' approach to local food system governance, based on strong social capital and collaboration. Clapp and Fuch's (2009) understanding of power also sheds light on the success of Correns's hybrid governance approach, based on three key factors: 1) mayoral leadership and the time and financial resources made available by the municipality (leveraging institutional power), 2) the development of a new citizen-based governance structure to ensure equal relationships of power between state and society (leveraging structural power), and 3) using organic agriculture as a mobilizing framework for sustainable community development (leveraging discursive power). It also stresses that the strong community bonds and the overlapping roles of members within smaller rural communities create vital opportunities for sustainable development, namely in the development of trust and reciprocity between community members.

This chapter is part of a series of four cases developed as part of my doctoral research, which sought to explore the governance mechanisms being developed between civil society, the state, and private sector actors to support sustainable local food system initiatives in Canada and the European Union (EU). The research presented in this chapter is based on ten semi-structured interviews and over a dozen additional informal exchanges with a diversity of food system actors in Correns during fieldwork in 2013 and 2014. Formal participants included municipal officials, farmers, community organizers, and local business owners, who shared their experiences and opinions on their community's governance arrangements and sustainable food system initiatives. As a means to give back to the community, informal activities conducted during my fieldwork included attending and volunteering during various community events, workshops, or talks, touring relevant sites of local development, as well as observing organizational meetings. These methods allowed me to understand the different dimensions of Correns's efforts and how local actors conceptualized their village's governance processes. Their words are directly translated in the results presented below.

The need for new food system governance: Alternative food systems for local realities

As described in the introductory chapter, contemporary food systems can be predominantly characterized by productivist and neoliberal models of food and agriculture that favour the industrialization, intensification, and financialization of agriculture and agri-food systems (Clapp, 2011; McMichael, 2004). These priorities are maintained by a limited number of global players – primarily transnational corporate actors and international regulators – who leverage their dominant positions to reinforce the status quo and decrease the autonomy and control of both producers and consumers over how food is produced and managed (IPES-Food, 2016). Yet, current food systems have also been undergoing a process of "massive change and contradiction" (Bernstein, 2015, p. 11). The "roll-out" of new regulatory mechanisms in light of decreased responsibility from the state has allowed for a concentration of power into the hands of an increasingly smaller number of private sector actors alongside new governance regimes reliant on civil society and state support.

Current trends not only point to increasing disjointedness within current food systems, but also stress the limitations of relying on overly simplistic dichotomies to understand them (e.g. local/global, neoliberal/alternative). A focus away from strict dichotomies allows us to move away from only understanding different food system dynamics as working in competition against one another, and instead enables us to see the spaces in which to attempt "something fundamentally different" (Ikerd, 2008, p. 78). It also becomes all the more important to consider the possible hybridities within current governance arrangements, not only to better describe them, but also to tease out opportunities for how power may be leveraged differently for food system change. As Correns's blended governance approach will demonstrate, it can be through a combination of integrated market and non-market interactions that marginalized actors can shape their food landscapes and reclaim their decision-making power within food systems.

In contrast to the motivations and practices of the dominant food model, the local alternative food systems described here are rooted in community and place. These food systems broadly value social and environmental stewardship alongside economic sustainability (Ikerd, 2008; DuPuis, 2006; Lyson, 2004; Hinrichs, 2003). They rely on decentralized governance, locally rooted solutions, and strong community relationships. More specifically, alternative food system scholars working at the local level have drawn on the notion of embeddedness to explain how markets may be re-embedded in certain values at the local level because of the smaller scale at which producers and consumers interact (Sage, 2003; Hinrichs, 2000). In other words, proximity and more localized scales of action enable the development of social capital, itself a fundamental building block to creating alternative foodscapes. Social capital is built through interactions among people, and is built on trust, shared norms and rules, reciprocity, and the creation and maintenance of networks. High levels of social capital within a community have been shown to have a clear correlation to economic and social development as well as the ability to collectively manage public resources (Pretty, 2002; Putnam et al, 1994).

Local spaces have often been identified as sites of resistance to larger regional, national, and global systems, as it is at this level that the outcomes of neoliberal arrangements are most brought to bear (Hendrickson & Heffernan, 2002; Friedman & McMichael, 1989). Scale can thus be leveraged as a means to empower certain actors and represents a strategic choice in and of itself (Born & Purcell, 2006). It should be noted, however, that Connelly et al (2011), Born and Purcell (2006), and others caution against the "fetishization" of the local (Connelly et al, 2011, p. 313), whereby local food systems are deemed the *de facto* sustainable and more democratic alternative. Certainly, overemphasizing any system by an idealized set of traits does not allow proper consideration of the opportunities and challenges they create.

Nevertheless, at the local level, municipalities are increasingly providing the space to put alternative food systems into practice by recognizing the power of endogenous models for change and the potential for local governance spaces to respond more immediately to citizen needs and interests. In the context of rural development, Marsden and Sonnino (2008, p. 403) argue that locally driven food system initiatives can serve as the key "proactive development tool to promote more sustainable economies of scope and synergy".

Lastly, rural alternative food systems often support and are supported by peasant agriculture.[2] We are witnessing a large and growing number of farmers worldwide self-identifying with "peasantry" for the struggle for autonomy from dominant systems, the cooperation with like partners, and the support of environmental stewardship the term has come to represent (van der Ploeg, 2009). Particularly relevant to the farmers and producers of Correns, conceptualizing oneself as "peasant" creates a particular understanding of one's role in shaping solutions. Overall, determining alternative food systems and new governance arrangements in Correns is intrinsically linked to the relationship between private and public, food and place, human dignity, and the ability to choose to be empowered rather than powerless in the face of change.

Correns: Rural renaissance through hybrid governance arrangements

To understand Correns, one must simply consider its motto, *"la sensation d'un privilège"* ("A feeling of privilege") to appreciate the sense of pride felt by its residents. The village made its mark in 1995 as France's "first organic village" – with 95 per cent of Correns' current production under organic agriculture and with the other 5 per cent either in transition or following the principles of *Protection Biologique Intégrée*[3] (integrated crop protection). Isolated from the South of France's densely populated, politically conservative, and wealthy beach towns, Correns is a small rural village in the region of Provence whose population only grows to 1,000 over the summer months. With over 90 per cent of its population involved in small-scale agriculture in some way,[4] most of Correns's locals still significantly identify with a "*paysan*" (peasant) lifestyle.[5]

Correns's recent history played a significant role not only in shaping the economic focus of its early transition years, but also the incremental change residents underwent to adopt a new paradigm in favour of sustainable community development. In the 1970s and 1980s, France saw a progressive depopulation of its countryside. Once a village of over 1,500 residents, Correns's population declined to below 450 over two decades leading to the closure of many local businesses (EHESS, 2015). By the mid-1990s, Correns could no longer support basic community amenities, including a *boulangerie*, a grocer, or a doctor. One olive oil producer recalls the desperation felt by many locals at the time: "some of us even told our children not to stay, not to be *paysans* anymore, not to work the land". At the time, a number of residents remembered older Corrençois saying, "When crows fly over Correns, they turn their heads to avoid seeing the misery".

Michaël Latz became mayor of Correns at the height of Correns's economic difficulties in 1995, and would serve as a key figure in mobilizing the village residents for the next 30 years. Having left a career as head of an agrochemical firm and having realized dominant agribusinesses' role in "manipulating the system at their will", Latz progressively convinced both his municipal team and the *paysans* to leverage organic agriculture as a way to reshape power dynamics within the food system and to give purpose and agency back to the local community. Latz – also a wine producer himself – encouraged the Vignerons de Correns, the village's wine cooperative, to transition to organic agriculture as a way of adding value to their craft. At the time, the wine cooperative provided not only a majority of the village's income, but also a good portion of its employment. Its success would give local *paysans* a reason to remain on the land, to regain pride in their profession, and not lose their local identity as a result of increasing buy-outs of land by larger neighbouring cooperatives.

Early and ongoing discussions between the mayoral team and local *paysans* established a tightly knit relationship between state and market actors. In this regard, however, it is critical to note the power of small communities to build and maintain strong social ties, as well as develop trust. As in many small communities, it is difficult to draw clear lines between municipal officials, civil society, and the private sector (represented largely here by farmers and small businesses) in Correns. Thus, while various efforts will be labelled as market-led, civil society-based or institutionally driven, the lack of distinction between formal and informal roles in tightly knit communities enables various actors to share identities in a way that a *paysan*, a business owner, or even the mayor is considered a citizen or community member as much as they are defined by their formal position in the community.

Initial conversations between the mayoral team and the *paysans* further determined that ensuring residents' economic livelihood was of primary concern. At the time, most local *paysans* agreed to transition to organic as a last ditch effort to improve their economic livelihoods. As shared by the head of Corren's wine cooperative, Fabien Mistre: "At the time, all of this didn't have much to do with moral or ethical convictions for us . . . Of course, we wanted to leave the land better off for our children, but it was about making a living first". Yet, what began as a change in farming practices quickly developed into a broader ideological project in favour of sustainability, despite insufficient economic pay-off after the first five years of transition. "Financially, [the transition to organic] still wasn't working out well for us", Mistre recalls, "but we were so morally and ecologically convinced that we tightened our belt and kept going!".

Latz, however, has always considered economic sustainability as inseparable from social or environmental concerns:

> When we think of sustainability, we tend to think of the environment first, then society, then the economy, but for us, the economic aspect has always been crucial. Ecologists might resent that, but they're wrong. We won't be able to have an ecological revolution if people can't make a living off of it . . . The values of sustainability only work if its successes touch on all three of its pillars. The economy is a critical area of engagement that allows us to be able to take on broader issues.

Value-adding and relying on market-based structures allowed Correns's producers to become economically viable over time: rather than focus solely on local distribution, the producers sought to gain opportunities by playing into higher-level marketing schemes. For example, the Vignerons de Correns have capitalized on both the AOC and organic labels. First, AOC labelling has allowed them to remain strong in the French market, where quality labelling remains consumers' primary concern. Second, the organic label has allowed the Vignerons to tap into the broader European market, where organic wines have greater traction than in France. In the future, and due to Correns's unique *terroir*, both the coop and the village ultimately hope to gain their own local designation of origin, similar to those already attained by other wine areas in the region (e.g. Cassis, Bandol).

Correns's economic success would certainly not have been possible without the broader enabling framework for organic agriculture in France. It is worth noting that France is the third biggest organic market in the world (behind the USA and Germany) and the second largest market in the EU (Willer & Lernoud, 2017). While local *paysans* stressed the economic hardships of the first few years of transition, they also noted being supported by subsidies from the EU, which grew significantly in the mid-1990s and continue to provide one of the strongest institutional frameworks for organic agriculture in the world (Larsson, 2015).

Correns's embrace of market-based approaches also remained deeply embedded in the cooperative model, and the values of equality and shared responsibility. Infusing community and cooperative values into Correns's approach has been key to its success. Beyond the clear buying power of cooperatives over smaller independent producers and the historical place of the cooperative model in Correns,[6] the model enables community-building and engenders greater trust between producers. Mistre insisted:

> We would never stray from the coop model. We work together, make wine together, market together, have meetings whenever we want . . . It's fantastic. It makes us want to pass this down to our kids . . . Coops are such an important model for communities that they have to stay. I will always fight for them.

In addition, making best use of both local and global markets, most of Correns's producers still rely on direct marketing as their primary distribution channel, and while neither wholly communitarian nor neoliberal, this approach has been prioritized to allow for greater civic engagement around food and stronger connections between local residents. Even beyond the local level, the Vignerons de Correns have insisted on selling the majority of their organic wine through direct marketing by attending events in-person across the country. In 2014, 80 per cent of the Vignerons' 7,500 hectolitres of wine were sold through direct marketing – yet another means, the members emphasized, to avoid going through intermediaries and multiply producer–consumer interactions and trust. Though some scholars have argued that organic labelling and global organic commodity chains have enabled the "distancing" of modern food systems by allowing consumers to rely on a label (Blay-Palmer, 2008; Kneen, 1993), the Vignerons have sought to leverage both the recognition and price premium of the organic label at the national and European scale, while maintaining direct marketing schemes to counteract the "facelessness" of modern systems. Drawing once again on system hybridities, the Vignerons both rely on conventional commodified exchanges, while also giving primacy to trust in quality, relationships, and place through direct exchange.

Hand-in-hand with market-based initiatives, large municipal investments also went into revitalizing Correns, drawing on their own institutional power to enable change. Correns's municipal government proactively engaged in laying down policy-based foundations for food system relocalization. Between 1995 and 2001, Latz's first term emphasized bringing back a series of new food-related employment opportunities to the village. This significant level of municipal backing has never simply been a matter of securing the village's economic future, but, according to Latz, of "recreating a real community". For example, the village council invested in public funds to buy a *boulangerie* and a grocer in the village centre, alongside a public call to hire local artisans to move in; both the *boulanger* and grocer selected were then heavily supported by public funds and social housing until their enterprises successfully took off. Today, the municipal authorities continue to back sustainable food projects: in partnership with Correns21 (see below), Correns has improved food in the municipal canteen, which now boasts 59 per cent organic and 30 per cent local ingredients, paired with educational campaigns on food for school children (Correns, 2018).

Due to the opportunities create by a market-driven approach backed by state support during initial phases of action, Correns has encouraged a new wave of alternative food entrepreneurialism and the creation of local niche markets; but perhaps most importantly, these efforts have allowed agriculture to attract a number of young entrepreneurs, who now reconsider farming a viable profession. Since the mid-1990s, the average Vignerons coop member's age has dropped to between 30 and 40, dramatically lower than the national average farming age of 60. In recent years, the village has attracted a goatherd – who also helps maintain public forest spaces through controlled grazing, a grain producer, a number of horticulturalists, an apiarist, a chicken farmer ,and organic grocery cooperative. The growing number of young producers who have returned to the village spoke with a strong sense of pride in being able to make a living from preserving *paysan* agriculture in their hometown. In short, ensuring economic viability remains a key entry point for food system sustainability. When framed in the context of community development, market-led approaches are not separate from but embedded in socio-cultural value and can support the spark for rural revitalization.

As Correns progressively secured its financial future through organic agriculture, in 2008, civil society and municipal officials also began to discuss the possibility to bring further change to the structural power in Correns. As a result, the town created Correns21, a local chapter of the UN Agenda 21 sustainable development action plan based on collaborative governance.[7] As the first conversations were held between the mayoral team and residents who Latz knew would be interested in new local governance dynamics, an early challenge faced in developing collaborative governance schemes was in garnering the interest of a wider range of residents. However, residents quickly opened up to the idea of Correns21 for two primary reasons. First, those leading the Correns21 project held high levels of social capital within the community, particularly as a number of residents proposing the idea had already brought about positive changes to the town in its transition to organic agriculture. Second, the progressive change of residents' mindset can likely be explained by how others have sought to conceptualize organic farming not only as "value-based agriculture" (Darnhofer et al, 2011), but as a "frame-bridging" social movement that allows various interests to coincide (Larsson, 2015; Vandergeest, 2009; Sligh and Cierpka, 2007).

The adoption of organic practices bridges the ideological and the practical, and serves as a "mobilizing frame" (Larsson, 2015) for both those seeking to improve their economic livelihoods and those seeking broader change. Certainly in Correns, the adoption of organic farming practices due to economic interest grew into a broader framework for sustainable community development over time. As mentioned earlier by Mistre and Latz, the logic of organic agriculture served as an early placeholder for the sustainable community development goals that would ensue. A founding member of Correns21 further noted that local *paysans*' transition organic agricultural practices progressively sparked a growing number of formal and informal conversations about food democracy and food system sustainability overtime; and residents grew to share a joint ambition in seeing new project and plans through.

The goal of Correns21 was to allow the municipal council to work in horizontal collaboration with village residents, namely by creating a space for civil society to play the leading role in developing and implementing a new sustainable action plan for Correns. Though Correns had employed traditional participatory fora (including public consultations and resident surveys) in the past, the municipal team and interested residents discussed their shared desire to go beyond structures ultimately still relying on vertical democratic strategies, and bridge the gap between formal and informal decision-making spaces. Correns21 would provide the space for citizens to take greater ownership over the village's future through collective action.

As a result, Correns21 was first organized through a piloting committee of 15 volunteer members and a small monitoring committee which include Correns21's acting president, the mayor, and the deputy mayor. The group formed around UN's Agenda 21 guidelines for diversity, including gender balance, positions for both younger and older residents, *paysans*, business owners, and unemployed residents. Latz assured that no more than 20 per cent of Correns21's pilot committee would be made up of municipal councillors to ensure the body's independence. Many took to the idea of a local Agenda 21 scheme quite easily once explained, as it reminded a number of residents of a time when local democracy was more direct, when residents met frequently more to discuss the future of their village – and even of France – through open and informal conversations in the local cafe or village square. Just as Mistre described the importance of the cooperative model for the Vignerons and the town's market-based community development approach, one resident explained that Correns's community had always been about solidarity, making efforts towards participatory democracy a natural outcome of local life.

The ensuing informality in the institutionalization of Correns's collaborative governance arrangements is difficult to trace as both members of Correns21 and of the municipality indicated that most of their decisions and arrangements were initially made through informal conversations (a point also stressed by wine cooperative members and their preference to make big decisions in informal settings). In other words, civic engagement and new governance schemes become as much a social activity as a set of formalized procedures. Yet, it is from a place of mutual trust and shared identity as "citizen" that municipal officials and citizens were able to easily formalize their relationships under the Agenda 21 scheme.

In 2008, Correns21 first goals included creating an action plan over ten years following the completion of a comprehensive, inclusive, and cross-sectoral strategy, which included collection of survey data, public debates, and stakeholder interviews. The resulting action plan was defined along three main axes, each relating to specific priorities and actions defined by villagers. Between 2010 and 2014, each axis was implemented in partnership with the municipal government providing policy and financial support. Actions included the development of a community garden, major renovations to reduce energy consumption in public buildings and lighting, the creation of "nature at school" programs in partnership with the local school as part of students' regular curriculum, the construction of an Haute Qualité Environmentale[8] (HQE) intergenerational centre and municipal canteen, a Correns culture and culinary cookbook for residents and tourists, an ongoing village newsletter and event promotion to support community-building, and finally the creation of a community supported agriculture (CSA) scheme for the sustainable harvesting of firewood.

Overall, the collaborative governance approach of Correns21 – which now boasts 40 volunteers – and the municipal council has operated by using iterative processes that allow for regularly refined, improved, and adaptable strategies, alongside the market-based approaches that have maintained the town's economic vibrancy for over three decades. As mentioned by Correns21's previous head:

> We work both with [municipal council] and separate from them. We can just call when we need each other. When they have a project, they

> give it to us for advice, but we each keep to our roles, while reporting to each other on a regular basis. We keep our independence and they keep theirs as democratically elected officials.

Together, the two groups attempt to navigate when and where each of their roles and inputs are required without trying to overstep each other nor drag on deliberation processes.

By mobilizing a diversity of residents with a range of expertise, Correns has used the Agenda 21 scheme to create a critical space for deliberation between its residents. It has produced a more holistic and long-term vision of sustainable development, going beyond Correns's initial focus on changing its agricultural practices, by fostering greater dialogue between and amongst residents, and a more mindful evaluation of public policies by the municipal government. By 2014, two-thirds of the 19 actions originally laid out by Correns21 in 2008 had been achieved either through civil society initiatives, government-led programs, or in collaboration.

The participatory dynamics engendered at the civil society level also served to spark further action within the municipal government. To create greater consistency between the Corren21 plan and the municipality, Latz was inspired by the Correns21 dynamic and has since sought to make the municipal council more representative of village realities. Today, *vignerons, paysans*, artisans, business owners, school board members, as well as public and private sector retirees serve as Correns's municipal council; its average age is 37 years, while the average age of many of the councils in surrounding villages is closer to 60. The goal of a diverse council was partially to demonstrate the municipal government's support for the *paysan* way of life and to stress that council values are not about "having large land-owners and producers bully our decisions", as emphasized by Latz.

Certainly, one should take note of the power of social conformity that can be particularly strong in smaller community. Part of Correns's success can be attributed to the strong motivator for locals to believe they should think and act in similar ways so as not to be singled out. As mentioned by one cooperative member, "We're a small village, so we all know each other. Of course, this motivates everyone to be a little more aware of what we're doing, because we're constantly watching what the others are doing". While this characteristic can strengthen a town's collective resolve, it may also subsume dissenting views which may exist within the town. Indeed, no criticism of Correns's trajectory was ever shared by formal interview or in informal conversations, in part because of the communal solidarity between the Corrençois – especially when discussing local dynamics with "outsiders". This was particularly evident when conducting research, during which village residents clearly seemed to try to steer discussion towards the positive aspects of Correns's efforts rather than to bring up any contention. This was striking in contrast to my own research experiences in other municipalities, where the interview process in municipalities that did not benefit from equally high levels of bonding capital occasionally became an outlet to share criticisms and weaknesses of one's community (Clément, 2017).

Putting one's eggs in many (complementary) baskets: Hybrid governance approaches for food system change

While rural areas across the globe are undergoing a "silent blight" (Tisdall, 2015), Correns's future seems all but secure. Just over two decades since Correns's foray into organic agricultural practice, a culture of sustainability has become undeniably ingrained within the community. The case of Correns and the strength of social capital within the village first allow us to reconsider how alternative values may be embedded into market-based governance arrangements. Following a Polanyian logic, capitalist society has commodified labour and land, creating a relationship between economy and community under the dominant organizing principle of self-gain over integration (Clark, 2014, p. 64–65). Polanyi emphasized how "a market society" becomes one stripped of the necessary characteristics required to function as a human society. In other words, Polanyi's work was not a critique of the economy but of morality. Societal transformation would thus lie in the ability to harness modern industrial society and translate its priorities of economic stability and distribution into "relational and collective freedom organized around the reproduction of social and ecological bonds" (Clark, 2014, p. 65). Put simply, the economy should serve the double function of fulfilling material needs while reproducing the necessary bonds of community.

Considering alternative food system schemes in light of the "Polanyian problem" (i.e. that a free market economy and a strong and stable society cannot co-exist) is to acknowledge that the primary issue with our current dominant food system is that it creates and perpetuates a market society (i.e. a society embodying and "enslaved" to capitalist market thinking) (Clark, 2014). In contrast, the *vignerons* of Correns use of the organic label to penetrate new markets has allowed generally marginalized actors to leverage modern market mechanisms in ways that value the norms and beliefs behind their practices (e.g. quality, place, cooperative management, direct contact with consumers).

Further, organic agriculture served as Correns's entry point into a broader sustainable development movement that allows community needs to supersede "the capitalist principle of self-gain and its market pattern" (Clark, 2014, p. 65). This case also highlights that local control of resources and collaboration neither implies a desire to end globalization nor to pursue local self-sufficiency. Instead, communities like Correns take pride in both exporting their own products, as well as showcasing those of other regions in their own supermarkets and through community events.

Similarly, while organic certification is a nationally and even globally recognized certification, global designations were seen as opportunities for small or traditional producers and cooperatives to market themselves in ways that satisfy new demands beyond their local market. To a community's advantage, leveraging the values of local production methods or of the particular agricultural advantages of a specific place (e.g. through the notion of terroir), "local" becomes both a process and a product, just as "organic" was understood as both a practice and the jumping off point to a broader philosophy.

The artificial dichotomy between globalized market-based and local community-centred systems must be reconsidered here. Correns's economy allows local producers to benefit from local, national, and regional markets to ensure their economic sustainability. The ability to then export these local products regionally, nationally, or even internationally were never called into question. Instead, the export of local product and knowledge was viewed as a positive means to scale out and connect with other like-minded communities and citizens around the globe, while providing the financials means to support ongoing and new local food system initiatives. Certainly, beyond the economic gains these arrangements may bring, they have also enabled citizens to feel renewed pride for the land to which they belong. "Belonging" remains a core sentiment in many rural areas, yet this tradition is losing ground as rural areas are progressively abandoned (EESC, 2017).

Though initially based on economic revitalization, Correns's long-term governance approach hinges on the desire to take control back from a larger system, whose processes serve to hold them at a disadvantage. Correns's story begun as one in which the village had reached an extreme, in which marginalization and loss experienced over decades propelled residents to consider new ways of structuring their economy and their agricultural practices through the market. Yet the hybrid governance approach that ensued has been far more ambitious, 1) by reclaiming power from dominant agrochemical companies and commodity chain intermediaries by shifting to organic agricultural practices, 2) gaining a renewed sense of identity when confronted with the placeless and faceless of current food system by adopting direct marketing strategies and the AOC place-based quality label, and 3) restructuring decision-making powers through collaborative governance schemes by developing Correns21, an arrangement under which the state has not been an obstacle to civil society, but rather, as the instrumental power enabling Correns's sustainable development.

Correns also highlights the strength of hybrid governance, in allowing each local actor to play to their strengths. Market-based governance schemes enable the local *vignerons* to benefit from global organic markets, while allowing other local *paysans* to produce organic produce for direct local consumption. Early uses of informal collaborative decision-making allowed municipal officials and the local *vignerons* to make critical decisions on the community's economic future by transitioning to organic. This ongoing state–society collaboration would later enable the creation of Correns21, a body existing outside of formal municipal institutions, yet with some decision-making power over the community's development.

The collaborative governance schemes as adopted in Correns proved successful not only when the interests and perspectives of *paysans*, municipal officials, entrepreneurs, and civil society actors were brought together, but also when taking into account the needs of different demographic groups including both men and women, young and old, and residents from a range of socio-economic backgrounds. This further allowed citizens to view themselves as more than consumers or passive citizens, but to understand their place and their power in such a way that people, the economy, and natural environments are consciously and deeply intertwined. Here, the overlapping roles of civil society, market actors, and municipal officials and the community bonds found in a small local community were key enabling factors to enabling this shift.

The establishment of Correns21 also touches on elements of deeper collaborative and perhaps polycentric governance, by setting new institutional and informal norms for decision-making based on community self-governance. However, the significant role of the state and the overlapping roles of residents in small rural communities (as market, social, and political actors) blur the possibility for a clearly demarcated space on the governance continuum.

Allowing for hybrid food governance schemes gives greater primacy to promoting local traditions and culture in ways which are not always valued by mainstream approaches. In other words, Correns has regained the "feeling of privilege" so crucial to maintain a resilient community by findings its purpose. Part of this privilege in Correns has been in the reclaiming the communal experience of the *paysan* lifestyle, and giving increased power, purpose, and relevance to the peasant voice; at times denoting extreme levels of poverty and hardship, at others highlighting a quiet strength and pride linking people to their land. The term itself speaks to a hybridity of purpose as a steward of the land, a political agent, a keeper of tradition and knowledge, as an economic actor.

Reflecting on power within the community, the municipal team leveraged its instrumental power through the technical and financial aid it provided to support civil society efforts established in the Correns21 action plan. Ongoing discussions between *paysans*, municipal officials, and other members of civil society drew on their shared discursive power to build on the bridging framework of organic to instill a new paradigm in favour of sustainable development in the community. The bridging frame of organic is found in the town's own attribution, "France's first organic village", meant not only to capture the set of practices it adopted, but the broader values and norms ascribed to the term by local residents, including understanding organic as the means to uphold a *paysan* way of life. Lastly, mutual recognition of each other's strengths enabled both the state and civil society to renegotiate their structural power, by establishing new rules and bodies determining who gets to sit at the decision-making table.

Hybrid food system arrangements can be driven by a series of complexities: hybridities of scale (local, regional, and national), of economic system (neoliberal and community-oriented), of production (organic and non-organic), and of decision-making (hierarchical and collaborative), amongst others. Blurring the lines between various governance strategies and arrangements allow us to imagine the wealth of possibilities they might engender, rather than the contradictions they might entail. Indeed, these dichotomies leave little room to consider where systems, strategies, and possibilities may intersect. This is not to downplay that governance arrangements should remain rooted in values and norms that explicitly seek to ensure that these systems remain accountable to their citizens, nor is this to say that sustainable food system initiatives do not acknowledge and seek to move away from the more negative effects of current trends in dominant food system structures, but rather, that such hybridities create the much-needed momentum for food system change.

Acknowledgements

Many thanks to the Centre for European Studies and the EU Commission for the Graduate Travel-Research Fellowship of the European Union that allowed me to complete my field work in the summer of 2014. A special thank you to all those in Correns who shared their experiences with me either through formal interview or informal conversation. This work would not have been possible without your enthusiasm and hospitality.

Notes

1 AOC translated as "Controlled Designation of Origin" is France's geographical indication certification granted to agricultural products based on the notion of *terroir* by the Ministry of Agriculture. *Terroir* approximately refers to the physical, physiological, and cultural factors that affect the quality of a crop or product. It not only refers to the specificities attribute to growing products in a certain soil, under certain weather conditions etc., but also encapsulates more intangible qualities such as local traditions, human knowledge and skills, and connection to the land.

2 Briefly defined, peasant agriculture can be characterized by 1) higher levels of intensive labour; 2) primacy on the environment (e.g. focus on soil and water health, use of polycultures vs. monocultures); 3) a culture of cooperation and community (e.g. knowledge and resource sharing vs. privatization of expertise and production); and relatedly, 4) a degree of independence from outside forces, relating also to marketing (e.g. use of direct marketing channels) (van der Ploeg, 2009).

3 *Protection Biologique Integrée* is a non-certified approach to agriculture similar to integrated pest management. It is adopted by a few local farmers who deem organic certification too costly or unnecessary when direct marketing their produce.

4 80 per cent of Correns is in wine production, of which 70 per cent is produced through its main cooperative, Les Vignerons de Correns. The remaining 30 per cent is produced by independent vineyards.

5 All participants used the term *paysans* to distinguish the town's producers from others who produce food and wine in the region. While directly translated to "peasant", the term captures the broader lifestyle involved in growing food, living in rural spaces, and strongly identifying with the land, and contrasts with terms more frequently associated with productivist agriculture.

6 Prior to the Vignerons de Correns, two cooperatives dominated the village, representing the republican–monarchist divide of the Napoleonic era, and later the local conservative–communist rift. Historically, members of each faction rarely spoke to each other and tensions between their families ran high (interview 4). It was only following the difficulties faced during the First World War that social differences were put aside to work under a single local cooperative, later renamed the Vignerons de Correns. Though decades have passed, residents still consider these stories as indicative of Correns's willingness to pursue sustainable food system development, emphasizing the historical trajectory that created the village's identity today.

7 Signed at the 1992 Rio Earth Summit, Agenda 21 features an entire section on the role of local political authorities in promoting sustainable development (see Section 28.1 and 28.3, UNEP, 1992).

8 HQE standards are France's green building standard overseen by the Association pour la Haute Qualité Environmental in Paris. They were derived from the principles of sustainable development set out at the 1992 UN Earth Summit.

References

Bernstein, H. (2015) *Food regimes and food regime analysis: a selective survey.* Presentation to the 'Land grabbing, conflict and agrarian-environmental transformations: perspectives from East and Southeast Asia' conference, 4–5 June 2015, Chiang Mai University, Thailand.

Blay-Palmer, A. (2008) *Food fears: from industrial to sustainable food systems.* New York, Routledge.

Born, B. & Purcell, M. (2006) Avoiding the local trap: scale and food systems in planning research. *Journal of Planning Education and Research*, 26, 195–207.

Clapp, J. (2011) *Food.* Cambridge, Polity.

Clapp, J. & Fuchs, D. (2009) Agrifood corporations, global governance, and sustainability: a framework for analysis. In: Clapp, J. & Fuchs, D. (eds) *Corporate power in global agri-food governance.* Cambridge, MIT Press, pp. 1–25.

Clark, T. D. (2014) Reclaiming Karl Polanyi, socialist intellectual. *Studies in Political Economy*, 94(1), 61–84.

Clément, C. W. (2017) *The new governance of sustainable food systems: shared insights from four rural communities in Canada and the EU.* Doctoral Thesis, Carleton University, Ottawa Canada.

Connelly, S., Markey, S., & Roseland, M. (2011) Bridging sustainability and the social economy: achieving community transformation through local food initiatives. *Critical Social Policy*, 31(2), 308–324.

Correns (2018) *Bulletin municipal n.37 – Printemps 2018.* Available at: http://cdn1_2.reseaudesvilles.fr/cities/126/documents/kftv6w9ket0lpaq.pdf

Darnhofer, I., Lindenthal, T., Bartel-Kratochvil, R., & Zollitsch, W. (2011) Conventionalisation of organic farming practices: from structural criteria towards an assessment based on organic principles. In: Lichtfouse, E., Hamelin, M., Navarrete, M., & Debaeke, P. (eds) *Sustainable Agriculture.* Volume 2. Dordrecht, Springer, pp. 331–349.

DuPuis, E. M. (2006) Civic markets: alternative value chain governance as civic engagement. *Crop Management*, 5(1). Doi:10.1094/CM-2006-0921-09-RV.

EESC (European Economic and Social Committee) (2017) *NAT/698 on villages and small towns as catalysts for rural development: challenges and opportunities.* Brussels, EESC Opinion.

EHESS (École des hautes études en sciences sociales) (2015) *Notice Communale.* Available at: http://cassini.ehess.fr/cassini/fr/html/fiche.php?select_resultat=10388

Friedman, H. & McMichael, P. (1989) Agriculture and the state system: the rise and decline of national agricultures, 1870 to the present. *Sociologia Ruralis*, 29(2), 93–117.

Hendrickson, M. K. & Heffernan, W. D. (2002) Opening spaces through relocalisation: locating potential resistance in the weaknesses of the global food system. *Sociologia Ruralis*, 42(4), 347–369.

Hinrichs, C. C. (2000) Embeddedness and local food systems: notes on two types of direct agricultural market. *Journal of Rural Studies*, 16(3), 295–303.

Hinrichs, C. C. (2003) The practice and politics of food system localization. *Journal of Rural Studies*, 19, 33–45.

Ikerd, J. (2008) *Crisis and opportunity: sustainability in American agriculture.* Lincoln, University of Nebraska Press.

IPES-Food (2016) *From uniformity to diversity: a paradigm shift from industrial agriculture to diversified agroecological systems.* Brussels, IPES-Food

Kneen, B. (1993) *From land to mouth: understanding the food System.* Toronto, NC Press.

Larsson, T. (2015) The rise of the organic foods movement as a transnational phenomenon. In: Herring, R. J. (ed.) *Food, Politics and Society.* Oxford, Oxford University Press, pp. 739–754.

Lyson, T. (2004) *Civic agriculture: reconnecting farm, food and community.* Hanover, Tufts University Press.

Marsden, T. & Sonnino, R. (2008) Rural development and the regional state: denying multifunctional agriculture in the UK. *Journal of Rural Studies*, 24(4), 422–431.

McMahon, M. (2014) Local food: food sovereignty or myth of alternative consumer sovereignty? In: Andrée, P., Ayres, J., Bosia, M. J., & Massicotte, M. J. (eds) *Globalization and food sovereignty: global and local change in the new politics of food.* Toronto, University of Toronto Press, pp. 111–138.

McMichael, P. (2004) *Development and social change: a global perspective.* Third edition. Thousand Oaks, Pine Forge Press.

Paquet, G. & Wilson, C. (2011) Collaborative co-governance as inquiring systems. *Optimum Online*, 41(2). Available at: www.optimumonline.ca/print.phtml?e=4JdoCAfcmUx22&id=492

Pretty, J. (2002) Social capital and the collective management of resources. *Science*, 302(5652), 1912–1914.

Putnam, R., Leonardi, R., & Nanetti, R. Y. (1994) *Making democracy work: civic traditions in modern Italy.* Princeton, Princeton University Press.

Sage, C. (2003) Social embeddedness and relations of regard: alternative "good food" networks in south-west Ireland. *Journal of Rural Studies*, 19(1), 47–60.

Sligh, M. & Cierpka, T. (2007) Organic values. In: Lockeretz, W. (ed.) *Organic farming: an international history.* Wallingford, CAB International, pp. 30–39.

Tisdall, S. (2015) Silent blight in a countryside of empty homes and shut shops. *The Guardian*, 23 August.

UNEP (United Nations Environment Programme) (1992) *Local Authorites' Initiatives in Support of Agenda 21.* Available at: www.un-documents.net/a21-28.htm

United Nations Habitat (2016) *World Cities Report 2016: Urbanization and development, emerging futures.* New York, United Nations Human Settlements Programme.

Vandergeest, P. (2009) *Opening the green box: how organic became the standard for alternative agriculture in Thailand*. Presentation at 'Berkeley Workshop on Environmental Politics'. April 17.
van der Ploeg, J. D. (2009) *The new peasantries: struggles for autonomy and sustainability in an era of empire and globalization*. London, Routledge.
Willer, H. & Lernoud, J. (eds) (2017) *The world of organic agriculture: statistics and emerging trends 2017*. Frick, Bio Suisse & FiBL.

Index

Page numbers in *italics* refer to figures; page numbers in **bold** refer to tables

For Product Safety Concerns and Information please contact our EU representative GPSR@taylorandfrancis.com
Taylor & Francis Verlag GmbH, Kaufingerstraße 24, 80331 München, Germany

www.ingramcontent.com/pod-product-compliance
Ingram Content Group UK Ltd.
Pitfield, Milton Keynes, MK11 3LW, UK
UKHW011010180926
475701UK00004B/661

* 9 7 8 0 3 6 7 6 6 1 8 2 3 *